DR. S.

Dr. B. TULASI LAKSHMI DEVI

Dr. G. SRINIVAS

G.V.P.N. SRIKANTH

FINITE ELEMENT ANALYSIS OF CONVECTIVE HEAT TRANSFER FLOW OF A MICROPOLAR FLUID THROUGH A POROUS MEDIUM IN CHANNELS/DUCTS

CANADIAN

Academic Publishing

2015

FINITE ELEMENT ANALYSIS OF CONVECTIVE HEAT TRANSFER FLOW OF A MICROPOLAR FLUID THROUGH A POROUS MEDIUM IN CHANNELS/DUCTS

Dr. S.SREENATHA REDDY
Principal & Professor
Guru Nanak Institute of Technology

Dr. B. Tulasi Lakshmi Devi
Associate Professor of Mathematics,
Guru Nanak Institute of Technology Hyderabad.

Dr. G. SRINIVAS
Professor, Department of Mathematics,
Guru Nanak Institute of Technology, Hyderabad.

G.V.P.N.Srikanth
Research Scholar in JNTUH, Hyderabad.

CANADIAN
Academic Publishing

2015

Price : $27.86

First Edition : 2015

ISBN : 978-1-926488-12-7

ISBN Allotment Agency : Library and Archives Canada (Govt. of Canada)

Published & Printed by
Canadian Academic Publishing
81, Woodlot Crescent,
Etobicoke,
Toronto, Ontario, Canada.
Postal Code- M9W 6T3
Phone- +1 (647) 633 9712
http://www.canadapublish.com

CONTENT

CHAPTER - I

FINITE ELEMENT ANALYSIS OF HYDROMAGNETIC CONVECTIVE HEAT TRANSFER FLOW OF A MICROPOLAR FLUID THROUGH A POROUS MEDIUM OVER A SEMI-INFINITE POROUS STRETCHING SHEET

1.1. INTRODUCTION:

The study of buoyancy driver convection flows through porous media has been stimulated by its application in several geophysical and engineering problems. Interests in understanding the connection transportation in porous material is increasing owing to the development of geothermal energy technology, high performance insulation for building and cold storage drying technology and many other areas.

The theory of micro fluids, as developed by Eringen [6] has been a field of active research for the last few decades as this class of fluids represents mathematically many industrially important fluids like paints, blood, body fluids, polymers, colloidal fluids and suspension fluids. In this material points in a volume element can undergo motions about centers of mass along with deformation. The problem of simple micro fluid contains a system of 19 equations with 19 unknowns so that it becomes difficult to find solution. A subclass of thee fluids introduced by Erignen [7], is the micro polar fluids, which exhibit the micro rotational effects and micro rotational inertia under these assumptions deformation of the fluid microelements is ignored: nevertheless micro rotational effects are still present and surface and body couples are permitted. Here in the skew symmetric property of the gyration tensor is imposed mathematically in addition to a condition of micro isotropy, so that the system of 19 equations reduces to seven equations in seven unknowns.

An excellent review of micro polar fluids and their applications was given by Arima et. al.[1]. Hoyt and Fabula [14] have shown experimentally that the fluids containing minute polymeric additives indicate considerable reduction of the skin (about 25. 30%) a concept which can be well explained by the theory of micro polar fluids. These fluids with microstructures are also capable of representing body fluids.

Power [19] has shown that the fluid flowing in brain (CSF) is adequately modeled by micro polar fluids.

This theory has been extended by Eringen [8] incorporating the thermal effects, i.e., heat dissipation, heat conduction to the so called as thermo micro polar fluids. Among the interesting results Eringen cited were the occurrence of a thermodynamic pressure tensor. The coupling of the temperature gradient with the constitutive equations and the occurrence of the micro rotation vector in the heat conduction equation. None of these effects were present in the classical field theories of fluids.

The free convective flow of the fluids with microstructure is an interesting area of research including liquid crystals, dilute solutions of polymers fluids and many types of suspensions, since in many configurations in technology and nature, one continually encounters masses of fluid rising freely in extensive effects Willaim et. al., [11] and Bhargava et. al., [4] investigated natural convection case the natural convection effects are also present because of the presence of gravitational body forces. A situation where both the natural and forced convection effects are of comparable order is called mined or combined convection. Mixed convection flows in channels and ducts and applications in nuclear reactors, heat exchangers etc., and have been studied by various authors namely Yucel [20], Gorla et. al.,[10]. Perhaps the most important question is the effect of buoyancy on forced convection transport rates. The buoyancy forces may aid or oppose forced flow causing an increase or decrease in the heat transfer rate. The problem of the stretching sheet has been of great use in engineering studies. Agarwal et. al., [3] studied the flow and heat transfer over a stretching sheet while Danborg and Fansler [9] have investigated a problem in which the free stream velocity is constant and the wall is being stretched with a variable velocity.

Recently Kelsan et. al., [14] studied the effect of surface conditions on the micro polar flow driven by a porous stretching sheet. The purpose of the present paper is to analyze the effect of surface conditions on mixed convection flow of a micro polar fluid driven by a porous stretching sheet, by assuming the most general type of boundary condition on the wall. Such a type of study may be applicable to polymer technology involving the stretching of the plastic sheet. In many metallurgical processes the cooling of continuous strips or filaments is involved by drawing them through a quiescent fluid. During the process of drawing, the strips are sometimes stretched. In such situations the rate of cooling has a great effect on the properties of the final product. By drawing them in a micro polar fluid the rate of cooling may be controlled, thereby giving the desired characteristics to the final product.

The set of coupled non-linear differential equations governing the flow, micro rotations and temperature fields were solved by using the finite element method and the results have been compared with those obtained by employing the quasi-linearization method. A discussion is provided for the effects of the Grashof number G, surface parameter s and suction parameter λ on the flow, micro rotation and temperature fields.

Many stages in nuclear reactors and MHD generators working under the influence of external magnetic fields could be examined and controlled using the present model. Na and Pop [17] investigated the boundary layer flow of a micro polar fluid past a stretching wall. Desscaun and Kelson [5] studied the flow of a micro polar fluid bounded by a stretching sheet. Hady [12] studied the solution of heat transfer to micro polar fluid from a non-isothermal stretching sheet with injection. In all the above studies, the authors took the stretching sheet to be oriented in horizontal direction. Abo-Eldahad and Ghnaim [2] investigated convective heat transfer in an electrically conducting micro polar fluid at a stretching surface with uniform free stream.

4

Mohammadein and Gorla [16] studied the heat transfer characteristics of a laminar boundary layer of a micro polar fluid over a linearly stretching sheet with prescribed uniform surface temperature or prescribed wall heat & flux and viscous dissipation and, internal heat generation. However, of late, the effects of a magnetic field on the micro polar fluid problem are very important. Mohammadein and Gorla [15] presented a numerical study for the boundary layer of a horizontal plate placed in a micro polar fluid. They analyzed the effects of a magnetic field with vectored surface mass transfer and induced buoyancy stream wise pressure gradients on heat transfer. They investigated the impact of the magnetic field, mass transfer, buoyancy, and material parameters on the surface friction and heat transfer rates. Siddheswar and Pranesh [19] investigated magneto-convection in a micro polar fluid.

In this chapter a finite element method solution for the fixed convection micro polar fluid flow through a porous medium driven by a porous stretching sheet with uniform suction under the influence of a uniform transverse magnetic field. The governing partial differential equations are solved numerically by using Galerkin finite element method analysis with quadratic approximation functions. The effect of porous medium, magnetic field and surface conduction on the velocity micro rotation functions has been studied. It is found that the micro polar fluids help in the reduction of stress also acts as a cooling agent.

1.2. FORMULATION OF THE PROBLEM:

Let us consider an isothermal, steady, Laminar, incompressible micro polar fluid through a porous medium flowing past a porous surface coinciding with the plane y = 0, the flow being confined in the region y > 0. Two equal and opposite forces are introduced along the x-axis so that the surface is stretched keeping the origin fixed. A uniform magnetic field of strength H_0 is applied normal to the walls. The component of velocity varies linearly along the x-axis. i.e., u(x, 0) = Dx where D (> 0) is an arbitrary constant. A uniform velocity V_0 through, and normal, to the stretching surface is also introduced. Let the wall temperature remains steady at T_w while the free stream fluid temperature remains steady at T_∞. Assuming the viscous dissipation effects to be negligible, the governing equations of the flow in two dimensions are as follows:

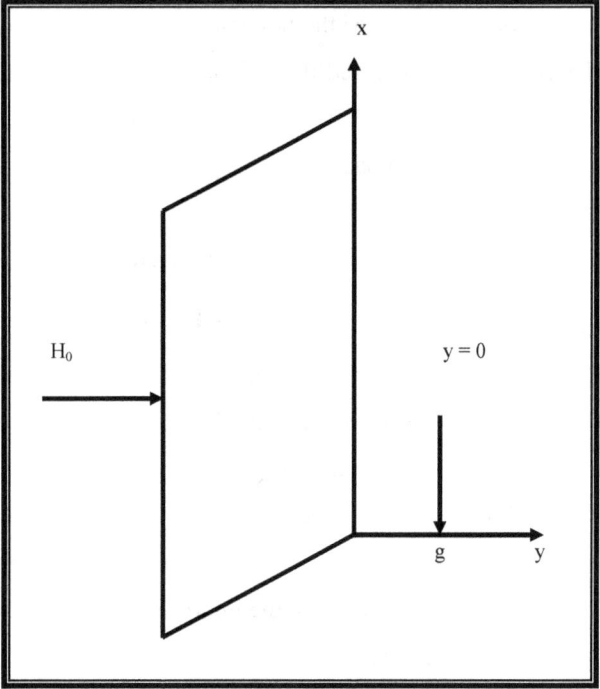

Mass:

$$\frac{\partial u}{\partial n} + \frac{\partial v}{\partial y} = 0 \qquad\qquad 2.1$$

Momentum:

$$u\frac{\partial u}{\partial x} + v\frac{\partial u}{\partial x} = \left(v + \frac{k}{\rho}\right)\frac{\partial^2 u}{\partial y^2} + \frac{k}{\rho}\frac{\partial N}{\partial y} + g_e\,\beta\,(T - T_\infty) - \left(\frac{\sigma\mu e^2 H_0^2}{\rho_0}\right)u - \left(\frac{v}{k}\right)u \qquad 2.2$$

Angular momentum:

$$u \frac{\partial u}{\partial x} + v \frac{\partial u}{\partial n} = -\frac{k}{\rho_j} \left(2N + \frac{\partial u}{\partial y} \right) + \frac{Vs}{\rho_j} \frac{\partial^2 N}{\partial y^2} \qquad 2.3$$

Energy equation:

$$\rho \, C_p \left(u \frac{\partial T}{\partial n} + v \frac{\partial T}{\partial y} \right) = k_f \frac{\partial^2 T}{\partial y^2} + \alpha \left(\frac{\partial T}{\partial n} \frac{\partial N}{\partial y} - \frac{\partial T}{\partial y} \frac{\partial N}{\partial x} \right) + Q \, (T_\infty - T) - \frac{\partial (u_r)}{\partial y} \qquad 2.4$$

where u, v are the velocity components along x and y directions, respectively, ρ is the fluid density, v is the kinematic viscosity, v_r is the kinematic rotational velocity, g is the acceleration of gravity, β_f is the coefficient of volumetric thermal expansion of the fluid. C_p is the specific heat at constant pressure, σ is fluid electrical conductivity, B is the magnetic induction, J is the micro inertia density, w is the component of the angular velocity vector normal to the x, y – plane, γ is the spin gradient viscosity, T is the temperature, α is the effective thermal diffusivity of the fluid and k is the effective thermal conductivity, q_r the radioactive heat flux. The second term on the right hand side of the momentum equation denotes buoyancy effects, and the third is the MHD term.

By using the Roseland approximation, the radioactive heat flux in the y direction is given by

$$q_r = \frac{-4\sigma_s}{3 \, ke} \frac{\partial T'^4}{\partial y}$$

Where σ_s Stefan-Boltzman constant and ke the mean absorption coefficient. It should be noted that by Roseland approximation we limit our analysis to optically thick fluids. If the temperature differences with in the flow are sufficiently small then equation can be linearized by expanding T^4 into the Taylor series about T_∞ and neglecting higher terms takes the form

$$T'^4 \cong 4T_\infty^3 T - 3 T_\infty^4.$$

The heat due to viscous dissipation is neglected for small velocities in energy equation. It is assumed that the porous plate moves with constant velocity in the longitudinal direction, and the plate temperature T varies exponentially with time. Under these assumptions, the appropriate boundary conditions velocity, micro rotation, and temperature fields are

at $y = 0 :$ $u(x,0) = Dx$, $v(x,\ 0) = V_0$, $N(x,) = -s\dfrac{\partial u}{\partial y}$, $T = T_w$

And as $y \to \infty$: $ui \to 0$, $N \to 0$, $T \to T_\infty$

Where $V_0 < 0$ represents the suction velocity while $V_0 > 0$ represents the injection velocity.

A linear relationship between the micro-rotation function N and the surface shear stress $\left(\dfrac{\partial u}{\partial y}\right)$ is chosen for investigating the effect to different surface conditions for the micro rotation. Here the boundary parameter and varies from 0 to 1. The first boundary condition (s = 0) is a generalization of the no slip condition, which requires that the fluid particles closet to a solid boundary stick to it-neither translating nor rotating. The second boundary condition i.e., micro rotation is equal to the fluid vorticity at the boundary (s $\neq$ 0) means that in the neighborhood of the boundary. The only rotation is due to fluid shear and therefore, the gyration vector must be equal to fluid vorticity.

If $\psi(x, y)$ represents the stream function,

Then $u = \dfrac{\partial \Psi}{\partial y}$, $v = \dfrac{-\partial \Psi}{\partial x}$ 2.5

Introducing the dimensionless functions $f(\eta)$ and $g(\eta)$ such that the continuity equation is automatically satisfied and by assuming

8

$$\eta = \left(\frac{D}{v(1+p_1)}\right)^{\frac{1}{2}} y, \quad \phi = (Dv(1+p_1))^{1/2} \times f(\eta) \qquad\qquad 2.6$$

$$N = \left(\frac{D^3}{v(1+p_1)}\right)^{\frac{1}{2}} \times g(\eta). \; \theta(\eta) = \left(\frac{T-T_\infty}{T_w - T_\infty}\right) \qquad\qquad 2.7$$

The governing equations (2.2) – (2.4) reduce to the following set of ordinary differential equations.

$$f'^2 - f f'' = f''' + \frac{p_1}{(1+p_1)} g' + G\theta(M^2 + D^{-1}) f' \qquad\qquad 2.8$$

$$f'g - f g' = \frac{-p_1}{p_3}(2g + f'') + \frac{p_2}{p_3(1+p_1)} g'' \qquad\qquad 2.9$$

$$\theta'' + p_r (f\theta' - \alpha g\theta') - \alpha\theta + \frac{4}{3 N_1}\theta'' = 0 \qquad\qquad 2.10$$

Where $p_1 = \dfrac{k}{\rho v}$, $p_2 = \dfrac{D v_s}{\rho v^2}$ and $p_3 = \dfrac{D_j}{v}$ are the physical micro polar parameters.

$$G = \left(\frac{g_e \, \beta \, (T_\omega - T_\infty)}{D^2 x}\right) \qquad \text{(Grashof Number)}$$

$$M^2 = \frac{\sigma \mu e^2 H_0^2 L^2}{v^2} \qquad \text{(Hartmann Number)}$$

$$D^{-1} = \frac{L^2}{k} \qquad \text{(Darcy Parameter)}$$

$$\alpha_1 = \frac{\theta L^2}{k f} \qquad \text{(Heat source parameter)}$$

$$N_1 = \frac{3\beta_r k_f}{4\sigma^\bullet Te^3} \qquad \text{(Radiation parameter)}$$

$$Pr = \left(\frac{v\rho C_p (1+p_1)}{k_f}\right) \qquad \text{(Prandtl number)}$$

9

and $\alpha = \left(\dfrac{\alpha' D}{\gamma \rho C_p (1+p_1)} \right)$ is the material parameter.

Using a rescaling of parameters as follows:

$$C_1 = \dfrac{p_1}{1+p_1}, \quad C_2 = \dfrac{p_2}{p_1(1+p_1)} \text{ and } C_3 = \dfrac{p_3}{(1+p_1)} \qquad 2.11$$

and introducing these parameter into equations (2.8) – (3.10) we get

$$f'^2 - f f'' = f''' + C_1 g' + G\theta + M_1^2 f' \qquad\qquad 2.12$$

$$C_2 g'' = 2g + f'' - \dfrac{C_3}{C_1}(f g' - f' g) \qquad\qquad 2.13$$

$$\theta'' + P_r (f \theta'' - \alpha g \theta') - \alpha_2 \theta = 0 \qquad\qquad 2.14$$

$$M_1^2 = M^2 + D^{-1}.$$

The corresponding boundary conditions then reduce to

$$f(0) = -\left(\sqrt{\dfrac{1-CD}{Dv}} \right), \quad v_0 = -\lambda, \quad f'(0) = 1,$$

$$g(0) = -s f''(0), \quad \theta(0) = 1 \text{ and } f'(\infty) = 0,$$

$$g(\infty) = 0 \text{ and } \theta(\infty) \to (0) = 0 \qquad\qquad 2.15$$

The local wall shear stress and wall couple stress may be written as

$$T_w = -(\mu + k)\dfrac{\partial u}{\partial y}\bigg|_{y=0} = -x f''(0)\sqrt{(\mu+k)D^3 \rho} \qquad 2.16$$

and $\quad m_w = v_s \dfrac{\partial N}{\partial y}\bigg|_{y=0} = \left(\dfrac{D^2 v_s}{v(1+p_1)} \right) x\, g'(0) \qquad\qquad 2.17$

The rate of heat transfer from the wall is given by

10

$$q_w = -k_f \left.\frac{\partial T}{\partial y}\right|_{y=0} = -k_f \ (T_w - T_\infty) \left(\frac{D}{v(1+p_1)}\right)^{\frac{1}{2}} \theta'(0) \qquad\qquad 2.18$$

1.3. METHOD OF SOLUTION:

Finite element method

To solve the differential equations (2.12 – 2.14) with the boundary conditions (2.15), we assume

$$f' = h \qquad\qquad 3.1$$

The system of equation (2.12 – 2.14) then reduces to

$$h'' + f h' - h^2 + c_1 g' + G\theta + (M^2 + D^{-1}) \ h \ = 0 \qquad\qquad 3.2$$

$$C_2 gh - h' - 2g + \frac{C_3}{C_1} (f g' - h g) = 0 \qquad\qquad 3.3$$

$$\theta'' + Pr \ (f\theta' - \alpha g\theta') - \alpha\theta = 0 \qquad\qquad 3.4$$

The corresponding boundary conditions now become

$$f(\theta) = -\lambda, \qquad h(0) = 1, \qquad g(0) = -sh'(0), \quad \theta(0) = 1,$$

$$h(\infty) = 0, \quad g(\infty) = 0, \qquad \theta(\infty) = 0 \qquad\qquad 3.5$$

For computational purposes and without any loss of generality, ∞ has been fixed as 8. The whole domain it divided into a set of 80 line elements of width 0.1. Each element being two nodded.

Implementation of the Finite Element Method

This method basically involves the following steps:

1) Division of the domain into linear elements, called the finite element mesh.
2) Generation of the element equations using variation formulations.

3) Assembly of element equations as obtained in step (2).

4) Imposition of the boundary conditions to the equations obtains in (3).

5) Solution of the assembled algebraic equations the assembled equations can be solved by any of the numerical technique viz., Gaussian elimination. LU decomposition method etc., The details of the steps given above can be found in Res (14)

Variational Formulation

The variational form associated with equations (3.1) – (3.4) over a typical two nodded linear element ($\eta_e \ \eta_{e+1}$)

$$\int_{\eta_e}^{\eta_{e+1}} w_1 (f' - h) \, d\eta = 0 \qquad\qquad 3.6$$

$$\int_{\eta_e}^{\eta_{e+1}} w_2 \, (h'' + f h' - h^2 + C_1 g' + G\theta + M_1^2 h) d\eta = 0 \qquad\qquad 3.7$$

$$\int_{\eta_e}^{\eta_{e+1}} w_3 \left[(C_2\theta'' - h' - 2\theta + \frac{C_3}{C_1} (f\theta' - h\theta) \right] d\eta = 0 \qquad\qquad 3.8$$

$$\int_{\eta_e}^{\eta_{e+1}} w_4 (N'' + Pr \, (f N' - \alpha\theta N' - \alpha N) \, d\eta = 0 \qquad\qquad 3.9$$

Finite Element Formulation

The finite element method may be obtained from (3.6) – (3.9) by substituting finite element approximations of the form

$$\left. \begin{aligned} f &= \sum_{k=1}^{3} f_k \Psi_k \\ h &= \sum_{k=1}^{3} h_k \Psi_k \\ \theta &= \sum_{k=1}^{3} \theta_k \Psi_k \\ N &= \sum_{k=1}^{3} N_k \Psi_k \end{aligned} \right\} \qquad\qquad 3.10$$

with $w_1 = w_2 = w_3 = w_u = \Psi_j^i \, (i, j = 1, 2, 3)$ \qquad\qquad 3.11

12

Using equation (2.9) we can write (3.6) – (3.9) as

$$\int_{\eta_e}^{\eta_{e+1}} \left(\frac{df}{d\eta} - h \right) \Psi_j^i \, d\eta = 0 \qquad\qquad 3.12$$

$$\int_{\eta_e}^{\eta_{e+1}} \left[\frac{d}{d\eta}\left(\frac{dh}{d\eta} \right) + f\frac{dh}{d\eta} + C_1\theta' + G\theta + M_1^2 h \right] d\eta = 0 \qquad\qquad 3.13$$

$$\int_{\eta_e}^{\eta_{e+1}} \left[C_2 \frac{d}{d\eta}\left(\frac{d\theta}{d\eta} \right) - \frac{dh}{d\eta} - 2\theta + \frac{C_3}{C_4}\left(f\frac{d\theta}{d\eta} - h\theta \right) \right] \Psi_j^i \, d\eta = 0 \qquad\qquad 3.14$$

$$\int_{\eta_e}^{\eta_{e+1}} \Psi_j^i \left[\frac{d}{d\eta}\left(\frac{dN}{d\eta} \right) + Pr\left(f\frac{dN}{d\eta} - 2\theta\frac{dN}{d\eta} - \alpha\, N \right) \right] d\eta = 0 \qquad\qquad 3.15$$

Following the Galarkin weighted residual method and integration by parts method to the equations (3.13)-(3.15) we obtain

$$\int_{\eta_e}^{\eta_{e+1}} \frac{d\Psi_j^i}{d\eta}\frac{dh}{d\eta}\, d\eta + \int_{\eta_e}^{\eta_{e+1}} f\frac{dh}{d\eta}\Psi_j^i \, d\eta + \int_{\eta_e}^{\eta_{e+1}} C_1\frac{d\theta}{d\eta}\Psi_j^i \, d\eta + \int_{\eta_e}^{\eta_{e+1}} G\theta\Psi_j^i \, d\eta$$
$$+ \int_{\eta_e}^{\eta_{e+1}} M_1^2 h\Psi_j^i \, d\eta = Q_{1,J} + Q_{2,J} \qquad\qquad 3.16$$

Where
$$-Q_{1,J} = \Psi_j(\eta_e)\frac{dh}{d\eta}(\eta_e)$$

$$Q_{2,J} = \Psi_j(\eta_{e+1})\frac{dh}{d\eta}(\eta_{e+1})$$

$$\int_{\eta_e}^{\eta_{e+1}} C_2\frac{d\Psi_j^i}{d\eta}\frac{d\theta}{d\eta}\, d\eta - \int_{\eta_e}^{\eta_{e+1}} \Psi_j^i\frac{dh}{d\eta}\, d\eta - 2\int_{\eta_e}^{\eta_{e+1}} \theta\Psi_j^i\frac{C_3}{C_1}$$
$$+ \int_{\eta_e}^{\eta_{e+1}} \left[f\Psi_j^i\frac{d\theta}{d\eta} - h\theta\Psi_j^i \right] = R_{1,J} + R_{2,J} \qquad\qquad 3.17$$

$$-R_{1,J} = \Psi_j(\eta_e)\frac{d\theta}{d\eta}(\eta_e) + \Psi_j(\eta_e)\frac{dh}{d\eta}(\eta_e)$$

$$R_{2,J} = \Psi_j(\eta_{e+1})\frac{d\theta}{d\eta}(\eta_{e+1}) + \Psi_j(\eta_{e+1})\frac{dh}{d\eta}(\eta_{e+1})$$

13

$$\int_{\eta_e}^{\eta_{e+1}} \frac{d\Psi_j^i}{d\eta} \frac{dN}{d\eta} d\eta + \Pr f \int_{\eta_e}^{\eta_{e+1}} f \Psi_j^i \frac{dN}{d\eta} d\eta - \int_{\eta_e}^{\eta_{e+1}} \alpha\theta\Psi_j^i \frac{dN}{d\eta} d\eta$$

$$- \int_{\eta_e}^{\eta_{e+1}} \alpha N \Psi_j^i d\eta = S_{1,J} + S_{2,J} \qquad 3.18$$

where

$$-S_{1,J} = \Psi_j(\eta_e) \frac{dN}{d\eta}(\eta_e) + \Psi_j(\eta_e) \frac{d\theta}{d\eta}(\eta_e)$$

$$S_{2,J} = \Psi_j(\eta_{e+1}) \frac{dN}{d\eta}(\eta_{e+1}) + \Psi_j(\eta_{e+1}) \frac{d\theta}{d\eta}(\eta_{e+1})$$

Expressing u^k, θ^k and N^k in terms of local nodal values in (3.16) – (3.18) we obtain

$$\sum_{k=1}^{3} h_k \int_{\eta_e}^{\eta_{e+1}} \frac{d\Psi_j^i}{d\eta} \frac{d\Psi_k}{d\eta} d\eta + \sum_{k=1}^{3} f_k h_k \int_{\eta_e}^{\eta_{e+1}} \frac{d\Psi_k}{d\eta} \Psi_j^i d\eta + C_1 \sum_{k=1}^{3} \theta_k \int_{\eta_e}^{\eta_{e+1}} \frac{d\Psi_k}{d\eta} \Psi_j^i d\eta +$$

$$\sum_{k=1}^{3} G\theta_k \int_{\eta_e}^{\eta_{e+1}} \Psi_k \Psi_j^i d\eta + \sum_{k=1}^{3} M_1^2 h_k \int_{\eta_e}^{\eta_{e+1}} \Psi_k \Psi_j^i d\eta = Q_{1,J} + Q_{2,J} \qquad 3.19$$

$$\sum_{k=1}^{3} \theta_k \int_{\eta_e}^{\eta_{e+1}} \frac{d\Psi_j^i}{d\eta} \frac{d\Psi_k}{d\eta} d\eta - \sum_{k=1}^{3} h_k \int_{\eta_e}^{\eta_{e+1}} \Psi_j^i \frac{d\Psi_k}{d\eta} d\eta - \sum_{k=1}^{3} 2\theta_k \int_{\eta_e}^{\eta_{e+1}} \Psi_k \Psi_j^i d\eta +$$

$$\sum_{k=1}^{3} \frac{C_3}{C_1} \theta_k \int_{\eta_e}^{\eta_{e+1}} f \Psi_j^i \frac{d\Psi_k}{d\eta} - \sum_{k=1}^{3} h\theta\Psi_j^i d\eta = R_{1,J} + R_{2,J} \qquad 3.20$$

$$\sum_{k=1}^{3} N_k \int_{\eta_e}^{\eta_{e+1}} \frac{d\Psi_j^i}{d\eta} \frac{d\Psi_k}{d\eta} d\eta + \sum_{k=1}^{3} \Pr f N_k \int_{\eta_e}^{\eta_{e+1}} f \Psi_j^i \frac{d\Psi_k}{d\eta} d\eta - \sum_{k=1}^{3} N_k \int_{\eta_e}^{\eta_{e+1}} \alpha\theta\Psi_j^i \frac{d\Psi_k}{d\eta} d\eta$$

$$- \sum_{k=1}^{3} \int_{\eta_e}^{\eta_{e+1}} \alpha \Psi_k \Psi_j^i d\eta = S_{1,J} + S_{2,J} \qquad 3.21$$

Choosing different Ψ_j^i's corresponding to each element η_e in the equation yields a local stiffness mat n_x of order 3×3 in the form

$$\left(f_{ij}^k\right)\left(u_i^k\right) - gG\left(g_{ij}^k\right)\left(\theta_i^k + NC_i^k\right) + g D^{-1}\left(M_{ij}^k\right)\left(u_i^k\right) + \delta^2 D\left(n_{ij}^k\right)\left(d_i^k\right) = Q_{2j}^k + Q_{1j}^k \qquad 3.22$$

Likewise the equations (3.20) – (3.21 gives rise to stiffness matrices

$$\left(e_{ij}^k\right)\left(\theta_i^k\right) - \Pr \rho_u \left(t_{ij}^k\right)\left(u_i^k\right) = R_{2j}^k + R_{1j}^k \qquad 3.23$$

14

$$\left(l_{ij}^k\right)\left(N_i^k\right) - R_e \, Pr\left(t_{ij}^k\right)\left(u_i^k\right) = S_{2j}^k + S_{1j}^k \qquad\qquad 3.24$$

where

$\left(s_{ij}^k\right), \left(g_{ij}^k\right), \left(m_{ij}^k\right)\left(n_{ij}^k\right)\left(e_{ij}^k\right), \left(d_{ij}^k\right)$ and $\left(t_{ij}^k\right)$ are 3×3 matrices and $\left(Q_{2j}^k\right), \left(Q_{1j}^k\right), \left(R_{2j}^k\right)\left(R_{1j}^k\right)\left(S_{2j}^k\right)$ and $\left(S_{1j}^k\right)$ Are 3x1 column matrices and such stiffness matrices (3.22) – (3.24) in terms of local nodes in each element are assembled using inter element continuity and equilibrium conditions to obtain the coupled global matrices in terms of the global nodal values of h, f, θ & ϕ. In case we choose n quadratic elements then the global matrices are of order 2n+1. The ultimate coupled global matrices are solved to determine the unknown global nodal values of the velocity, temperature and concentration in fluid region. In solving these global matrices an iteration procedure has been adopted to include the boundary & effects in the porous region.

The shape functions corresponding to

$$\Psi_1^1 = \frac{(y-8)(y-16)}{128},\qquad\qquad \Psi_2^1 = \frac{(y-24)(y-32)}{128},$$

$$\Psi_3^1 = \frac{(y-40)(y-48)}{128},\qquad\qquad \Psi_1^2 = \frac{(y-4)(y-8)}{32},$$

$$\Psi_2^2 = \frac{(y-12)(y-16)}{32},\qquad\qquad \Psi_2^3 = \frac{(y-20)(y-24)}{32},$$

$$\Psi_1^3 = \frac{(3y-8)(3y-16)}{128},\qquad\qquad \Psi_2^3 = \frac{(3y-24)(3y-32)}{128},$$

$$\Psi_3^3 = \frac{(3y-40)(3y-48)}{128},\qquad\qquad \Psi_1^4 = \frac{(y-2)(y-4)}{8},$$

$$\Psi_2^4 = \frac{(y-6)(y-8)}{8},\qquad\qquad \Psi_3^4 = \frac{(y-10)(y-12)}{8},$$

15

$$\Psi_1^5 = \frac{(5y-8)(5y-16)}{128},$$
$$\Psi_2^5 = \frac{(5y-24)(5y-32)}{128},$$

$$\Psi_3^5 = \frac{(5y-40)(5y-48)}{128}$$

Stiffness matrices:

The global matrix for θ is

$$A_3\, X3 = B_3 \qquad\qquad\qquad\qquad 3.30$$

The global matrix for N is

$$A_4\, X4 = B_4 \qquad\qquad\qquad\qquad 3.31$$

The global matrix for u is

$$A_5\, X5 = B_5 \qquad\qquad\qquad\qquad 3.32$$

The global matrix for f is

$$A_6\, X6 = B_6 \qquad\qquad\qquad\qquad 3.32$$

where

$$A3 = \begin{pmatrix}
-1 & a_{12} & a_{13} & 0 & 0 & 0 & 0 & 0 & 0 & 0 & 0 \\
0 & a_{22} & a_{23} & 0 & 0 & 0 & 0 & 0 & 0 & 0 & 0 \\
0 & a_{32} & a_{33} & a_{34} & a_{35} & 0 & 0 & 0 & 0 & 0 & 0 \\
0 & 0 & a_{44} & a_{44} & a_{45} & 0 & 0 & 0 & 0 & 0 & 0 \\
0 & 0 & a_{53} & a_{54} & a_{55} & a_{56} & a_{57} & 0 & 0 & 0 & 0 \\
0 & 0 & 0 & 0 & a_{65} & a_{66} & a_{67} & 0 & 0 & 0 & 0 \\
0 & 0 & 0 & 0 & a_{75} & a_{76} & a_{77} & a_{78} & a_{79} & 0 & 0 \\
0 & 0 & 0 & 0 & 0 & 0 & a_{87} & a_{88} & a_{89} & 0 & 0 \\
0 & 0 & 0 & 0 & 0 & 0 & a_{97} & a_{98} & a_{99} & a_{910} & 0 \\
0 & 0 & 0 & 0 & 0 & 0 & 0 & 0 & a_{109} & a_{1010} & 0 \\
0 & 0 & 0 & 0 & 0 & 0 & 0 & 0 & a_{119} & a_{1110} & -1
\end{pmatrix}$$

$$A4 = \begin{pmatrix}
-1 & b_{12} & b_{13} & 0 & 0 & 0 & 0 & 0 & 0 & 0 & 0 \\
0 & b_{22} & b_{23} & 0 & 0 & 0 & 0 & 0 & 0 & 0 & 0 \\
0 & b_{32} & b_{33} & b_{34} & b_{35} & 0 & 0 & 0 & 0 & 0 & 0 \\
0 & 0 & b_{43} & b_{44} & b_{45} & 0 & 0 & 0 & 0 & 0 & 0 \\
0 & 0 & b_{53} & b_{54} & b_{55} & b_{56} & b_{57} & 0 & 0 & 0 & 0 \\
0 & 0 & 0 & 0 & b_{65} & b_{66} & b_{67} & 0 & 0 & 0 & 0 \\
0 & 0 & 0 & 0 & b_{75} & b_{76} & b_{77} & b_{78} & b_{79} & 0 & 0 \\
0 & 0 & 0 & 0 & 0 & 0 & b_{87} & b_{88} & b_{89} & 0 & 0 \\
0 & 0 & 0 & 0 & 0 & 0 & b_{97} & b_{98} & b_{99} & b_{910} & 0 \\
0 & 0 & 0 & 0 & 0 & 0 & 0 & 0 & b_{109} & b_{1010} & 0 \\
0 & 0 & 0 & 0 & 0 & 0 & 0 & 0 & b_{119} & b_{1110} & -1
\end{pmatrix}$$

$$A5 = \begin{pmatrix}
-1 & \frac{2}{3} & -\frac{1}{6} & 0 & 0 & 0 & 0 & 0 & 0 & 0 & 0 \\
0 & 0 & \frac{2}{3} & 0 & 0 & 0 & 0 & 0 & 0 & 0 & 0 \\
0 & -\frac{2}{3} & 0 & \frac{2}{3} & -\frac{1}{6} & 0 & 0 & 0 & 0 & 0 & 0 \\
0 & 0 & -\frac{2}{3} & 0 & \frac{2}{3} & 0 & 0 & 0 & 0 & 0 & 0 \\
0 & 0 & \frac{1}{6} & -\frac{2}{3} & 0 & \frac{2}{3} & -\frac{1}{6} & 0 & 0 & 0 & 0 \\
0 & 0 & 0 & 0 & -\frac{2}{3} & 0 & \frac{2}{3} & 0 & 0 & 0 & 0 \\
0 & 0 & 0 & 0 & \frac{1}{6} & -\frac{2}{3} & 0 & \frac{2}{3} & -\frac{1}{6} & 0 & 0 \\
0 & 0 & 0 & 0 & 0 & 0 & -\frac{2}{3} & 0 & \frac{2}{3} & 0 & 0 \\
0 & 0 & 0 & 0 & 0 & 0 & \frac{1}{6} & -\frac{2}{3} & 0 & \frac{2}{3} & -\frac{1}{6} \\
0 & 0 & 0 & 0 & 0 & 0 & 0 & 0 & -\frac{2}{3} & 0 & \frac{2}{3} \\
0 & 0 & 0 & 0 & 0 & 0 & 0 & 0 & \frac{1}{6} & -\frac{2}{3} & \frac{1}{2}
\end{pmatrix}$$

17

$$A6 = \begin{pmatrix}
-1 & c_{12} & c_{13} & 0 & 0 & 0 & 0 & 0 & 0 & 0 & 0 \\
0 & c_{22} & c_{23} & 0 & 0 & 0 & 0 & 0 & 0 & 0 & 0 \\
0 & c_{32} & c_{33} & c_{34} & c_{35} & 0 & 0 & 0 & 0 & 0 & 0 \\
0 & 0 & c_{43} & c_{44} & c_{45} & 0 & 0 & 0 & 0 & 0 & 0 \\
0 & 0 & c_{53} & c_{54} & c_{55} & c_{56} & c_{57} & 0 & 0 & 0 & 0 \\
0 & 0 & 0 & 0 & c_{65} & c_{66} & c_{67} & 0 & 0 & 0 & 0 \\
0 & 0 & 0 & 0 & c_{75} & c_{76} & c_{77} & c_{78} & c_{79} & 0 & 0 \\
0 & 0 & 0 & 0 & 0 & 0 & c_{87} & c_{88} & c_{89} & 0 & 0 \\
0 & 0 & 0 & 0 & 0 & 0 & c_{97} & c_{98} & c_{99} & c_{910} & 0 \\
0 & 0 & 0 & 0 & 0 & 0 & 0 & 0 & c_{109} & c_{1010} & 0 \\
0 & 0 & 0 & 0 & 0 & 0 & 0 & 0 & c_{119} & c_{1110} & -1
\end{pmatrix}$$

$$h_{i,j} = f_{ij} + \delta D^{-1} m_{i,j} + \delta^2 + n_{ij}$$

$$X_3 = \begin{bmatrix} g_1 \\ g_2 \\ g_3 \\ g_4 \\ g_5 \\ g_6 \\ g_7 \\ g_8 \\ g_9 \\ g_{10} \\ g_{11} \end{bmatrix} \quad X_4 = \begin{bmatrix} \theta_1 \\ \theta_2 \\ \theta_3 \\ \theta_4 \\ \theta_5 \\ \theta_6 \\ \theta_7 \\ \theta_8 \\ \theta_9 \\ \theta_{10} \\ \theta_{11} \end{bmatrix} \quad X_5 = \begin{bmatrix} h_1 \\ h_2 \\ h_3 \\ h_4 \\ h_5 \\ h_6 \\ h_7 \\ h_8 \\ h_9 \\ h_{10} \\ h_{11} \end{bmatrix} \quad X_6 = \begin{bmatrix} f_1 \\ f_2 \\ f_3 \\ f_4 \\ f_5 \\ f_6 \\ f_7 \\ f_8 \\ f_9 \\ f_{10} \\ f_{11} \end{bmatrix}$$

$$B_3 = \begin{pmatrix} d_{11} \\ d_{21} \\ d_{31} \\ d_{41} \\ d_{51} \\ d_{61} \\ d_{71} \\ d_{81} \\ d_{91} \\ d_{101} \\ d_{111} \end{pmatrix}$$

$$B_4 = \begin{pmatrix}
\dfrac{-14c_1\left(-75s+100sh_2-25sh_3-\dfrac{5\lambda s(8s^2-500c_2)\left(-3s+4sh_2-sh_3\right)}{s^2}\right)+sc_3+\left(\dfrac{195\lambda s\left(-3s+4sh_2-sh_3\right)}{s}-\dfrac{350\lambda sf\left(-3s+4sh_2-sh_3\right)}{s^2}+\dfrac{100\lambda sh\left(-3s+4sh_2-sh_3\right)}{s}\right)}{2100sc_1} \\[4ex]
-\dfrac{-14c_1\left(-25s+25sh_3-\dfrac{5\lambda s(s^2-500c_2)\left(-3s+4sh_2-sh_3\right)}{s^2}\right)+sc_3+\left(\dfrac{25\lambda s\left(-3s+4sh_2-sh_3\right)}{s}-\dfrac{175\lambda sf\left(-3s+4sh_2-sh_3\right)}{s^2}+\dfrac{20\lambda sh\left(-3s+4sh_2-sh_3\right)}{s}\right)}{525sc_1} \\[4ex]
c_{31} \\[2ex]
-\dfrac{-14c_1\left(-25sh_3+25sh_4\right)}{525sc_1} \\[3ex]
-\dfrac{14c_1\left(-25sh_3+100sh_4-75sh_5\right)+sc_3}{2100sc_1}-\dfrac{-14c_1\left(-75sh_5+100sh_6-25sh_7\right)}{2100sc_1} \\[3ex]
-\dfrac{14c_1\left(-25sh_5+25sh_7\right)}{525sc_1} \\[3ex]
-\dfrac{14c_1\left(-25sh_5+100sh_6-75sh_7\right)+}{2100sc_1}-\dfrac{-14c_1\left(-75sh_7+100sh_8-25sh_9\right)}{2100sc_1} \\[3ex]
-\dfrac{-14c_1\left(-25sh_7+25sh_9\right)}{525sc_1} \\[3ex]
-\dfrac{14c_1\left(-25sh_7+100sh_8-75sh_9\right)}{2100sc_1}-\dfrac{14c_1\left(-75sh_9+100sh_{10}\right)}{2100sc_1} \\[3ex]
-\dfrac{-14c_1\left(-25sh_9\right)}{525sc_1} \\[3ex]
-\dfrac{14c_1\left(-25sh_9+100sh_{10}\right)}{2100sc_1}
\end{pmatrix}$$

$$B_5 = \begin{pmatrix}
-\dfrac{1}{150}\left(-4s+75\lambda-2sh_2+sh_3\right) \\[2ex]
\dfrac{1}{75}\left(-50\lambda+s\left(1+8h_2+h_3\right)\right) \\[2ex]
-\dfrac{1}{150}\left(s-25\lambda-2sh_2-4sh_3\right)-\dfrac{1}{150}\left(-4sh_3-2sh_4+sh_5\right) \\[2ex]
\dfrac{1}{75}\left(+s\left(h_3+8h_4+h_5\right)\right) \\[2ex]
-\dfrac{1}{150}\left(sh_3-2sh_4-4sh_5\right)-\dfrac{1}{150}\left(-4sh_5-2sh_6+sh_7\right) \\[2ex]
\dfrac{1}{75}\left(s\left(h_5+8h_6+h_7\right)\right) \\[2ex]
-\dfrac{1}{150}\left(sh_5-2sh_6-4sh_7\right)-\dfrac{1}{150}\left(-4sh_7-2sh_8+sh_9\right) \\[2ex]
\dfrac{1}{75}\left(s\left(h_7+8h_8+h_9\right)\right) \\[2ex]
-\dfrac{1}{150}\left(sh_7-2sh_8-4sh_9\right)-\dfrac{1}{150}\left(-4sh_9-2sh_{10}\right) \\[2ex]
\dfrac{1}{75}\left(s\left(h_9+8h_{10}\right)\right) \\[2ex]
-\dfrac{1}{150}\left(sh_9-2sh_{10}\right)
\end{pmatrix}$$

19

$$
B_6 =
\begin{cases}
\dfrac{7000-s\left(39\,s-56\,Gs-700\,\lambda+420\,\underset{2}{f}-70\,\underset{3}{f}+350\,c_1\left(\frac{45\ \lambda s}{s}-4\,\underset{2}{g+g}\right)-28\,Gs\,\underset{2}{\varrho}+14\,Gs\,\underset{3}{\varrho}\right)}{2100\,s} \\[2ex]
\dfrac{7000.\ -5\,s^2+s\left(105\,\lambda-280\,\underset{2}{f}+35\,\underset{3}{f}\right)-350\,s\,c_1\left(\frac{15\ \lambda s}{s}-\underset{2}{g}\right)+s\left(s\left(7G\left(1+8\,\underset{2}{\varrho}+\underset{3}{\varrho}\right)\right)\right)}{525\,s} \\[3ex]
\dfrac{3\,s^2+2\left(3500+s\left(-35\,\lambda+70\,\underset{2}{f}+70\,\underset{3}{f}\right)\right)+350\,s\,c_1\left(\frac{15\ \lambda s}{s}-4\,\underset{2}{g}+3\,\underset{3}{g}\right)-s\left(s\,14G\left(1-2\,\underset{2}{\varrho}-4\,\underset{3}{\varrho}\right)\right)}{2100\,s}\quad -s\left(350\,c_1\left(3\,\underset{3}{g}-4\,\underset{4}{g+g}\right)-56\,Gs\,\underset{3}{\varrho}-28\,Gs\,\underset{4}{\varrho}+14\,Gs\,\underset{5}{\varrho}\right) \\[3ex]
\dfrac{7000+s\left(-105\,\underset{3}{f}-280\,\underset{4}{f}+35\,\underset{5}{f}\right)-350\,s\,c_1\left(\underset{3}{g}-\underset{5}{g}\right)+s\left(s\left(7G\left(\underset{3}{\varrho}+8\,\underset{4}{\varrho}+\underset{5}{\varrho}\right)\right)\right)}{525\,s} \\[3ex]
\dfrac{350\,s\,c_1\left(\underset{3}{g}-4\,\underset{4}{g}+3\,\underset{5}{g}\right)-s\left(+s\left(14G\left(\underset{3}{\varrho}-2\,\underset{4}{\varrho}-4\,\underset{5}{\varrho}\right)\right)\right)}{2100\,s}\quad-\dfrac{\left(350\,c_1\left(3\,\underset{5}{g}-4\,\underset{6}{g}+g\right)-56\,Gs\,\underset{5}{\varrho}-28\,Gs\,\underset{6}{\varrho}+14\,Gs\,\underset{7}{\varrho}\right)}{2100\,s} \\[3ex]
\dfrac{\left(7000+s\left(-105\,\underset{5}{f}-280\,\underset{6}{f}+35\,\underset{7}{f}\right)-350\,s\,c_1\left(\underset{5}{g}-\underset{7}{g}\right)+s\left(s\left(7G\left(\underset{5}{\varrho}+8\,\underset{6}{\varrho}+\underset{7}{\varrho}\right)\right)\right)\right)}{525\,s} \\[3ex]
\dfrac{-s\left(s\left(14G\left(\underset{6}{\varrho}-2\,\underset{7}{\varrho}-4\,\underset{8}{\varrho}\right)\right)\right)}{2100\,s}\quad-\dfrac{\left(350\,c_1\left(3\,\underset{7}{g}-4\,\underset{8}{g+g}\right)-56\,Gs\,\underset{7}{\varrho}-28\,Gs\,\underset{8}{\varrho}+14\,Gs\,\underset{9}{\varrho}\right)}{2100\,s} \\[3ex]
\dfrac{\left(7000+s\left(-105\,\underset{7}{f}-280\,\underset{8}{f}+35\,\underset{9}{f}\right)-350\,s\,c_1\left(\underset{7}{g}-\underset{9}{g}\right)+s\left(s\left(+7G\left(\underset{7}{\varrho}+8\,\underset{8}{\varrho}+\underset{9}{\varrho}\right)\right)\right)\right)}{525\,s} \\[3ex]
\dfrac{350\,s\,c_1\left(\underset{7}{g}-4\,\underset{8}{g}+3\,\underset{9}{g}\right)-s\left(+s\left(+14G\left(\underset{7}{\varrho}-2\,\underset{8}{\varrho}-4\,\underset{9}{\varrho}\right)\right)\right)}{2100\,s}\quad-\dfrac{\left(350\,c_1\left(3\,\underset{9}{g}-4\,\underset{10}{g}\right)-56\,Gs\,\underset{9}{\varrho}-28\,Gs\,\underset{10}{\varrho}\right)}{2100\,s} \\[3ex]
\dfrac{\left(7000+s\left(-105\,\underset{9}{f}-280\,\underset{10}{f}+35\,\underset{11}{f}\right)-350\,s\,c_1+s\left(s\left(7G\left(\underset{9}{\varrho}+8\,\underset{10}{\varrho}\right)\right)\right)\right)}{525\,s} \\[3ex]
\dfrac{\left(3500+s\left(35\,\underset{9}{f}+70\,\underset{10}{f}+70\,\underset{11}{f}\right)\right)+350\,s\,c_1\left(-4\,\underset{10}{g}\right)-s\left(+s\left(14G\left(+\underset{9}{\varrho}-2\,\underset{10}{\varrho}\right)\right)\right)}{2100\,s}
\end{cases}
$$

The equilibrium conditions are

$$R_3^1 + R_1^2 = 0, \qquad\qquad\qquad R_3^2 + R_1^3 = 0$$

$$R_3^3 + R_1^4 = 0, \qquad\qquad\qquad R_3^4 + R_1^5 = 0$$

$$Q_3^1 + Q_1^2 = 0, \qquad\qquad\qquad Q_3^2 + Q_1^3 = 0$$

$$Q_3^3 + Q_1^4 = 0, \qquad\qquad\qquad Q_3^4 + Q_1^5 = 0$$

$$S_3^1 + S_1^2 = 0, \qquad\qquad\qquad S_3^2 + S_1^3 = 0$$

$$S_3^3 + S_1^4 = 0, \qquad\qquad\qquad S_3^4 + S_1^5 = 0 \qquad 3.33$$

Solving these coupled global matrices for temperature, concentration and velocity (3.3.-3.32) respectively and using the iteration procedure we determine the unknown global nodes through which the temperature, concentration and velocity of different radial intervals at any arbitrary axial cross section are obtained.

20

1.4. DISCUSSION:

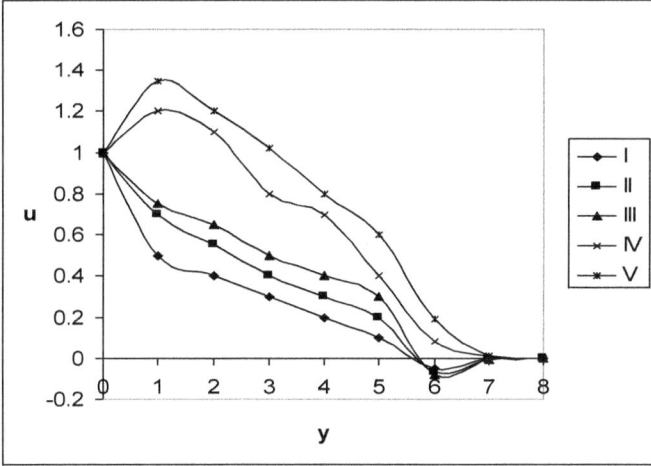

Fig. 1 : Variation of u with G

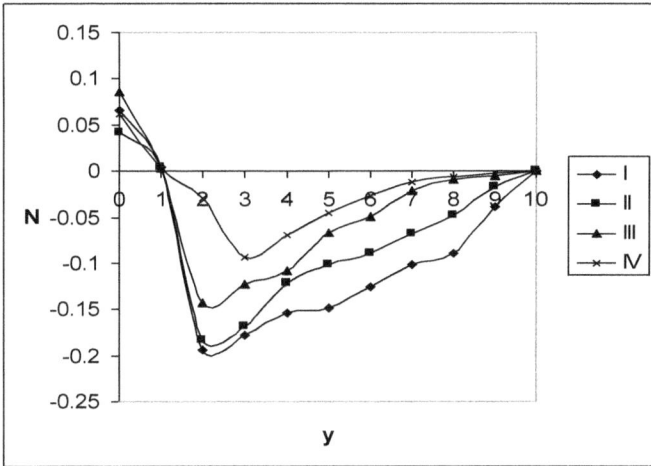

Fig. 2 : Variation of N with G

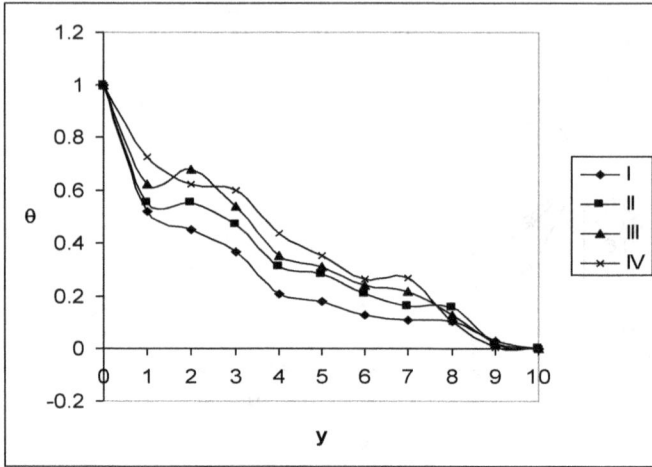

Fig. 3 : Variation of θ with G

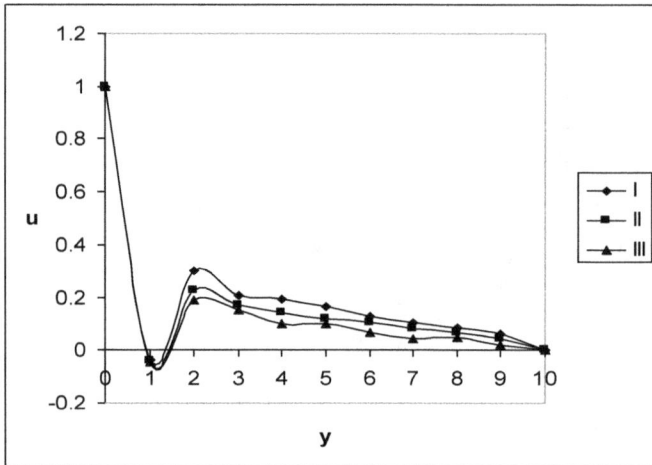

Fig. 4 : Variation of u with D⁻¹

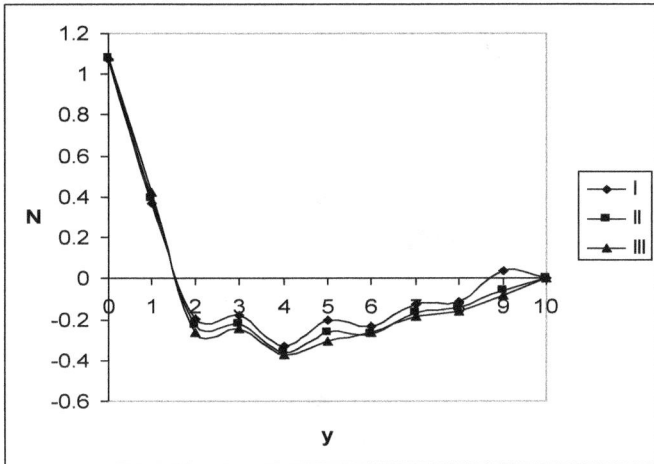

Fig. 5 : Variation of N with D^{-1}

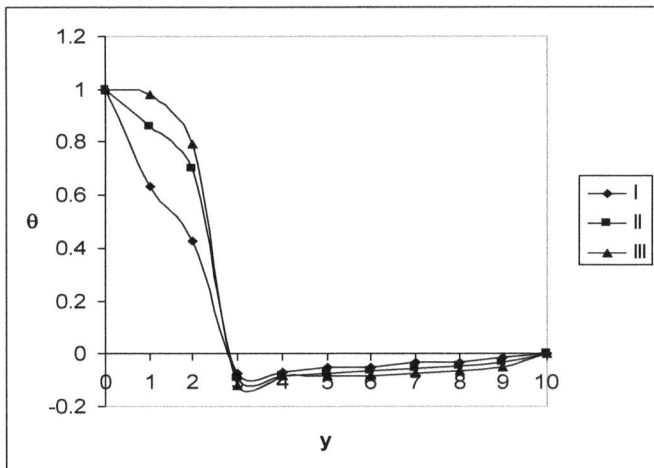

Fig. 6 : Variation of θ with D^{-1}

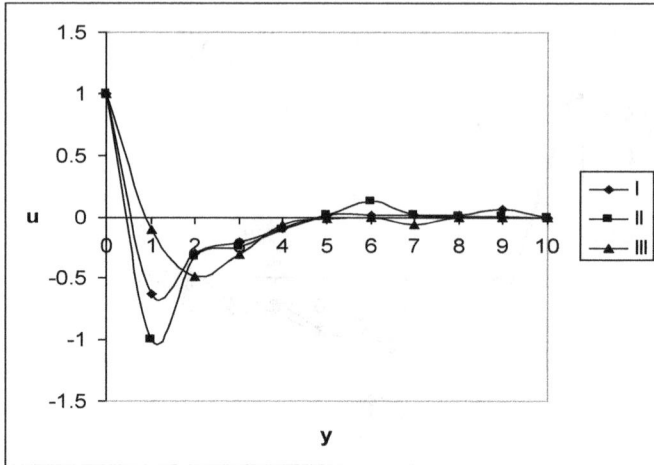

Fig. 7 : Variation of u with M

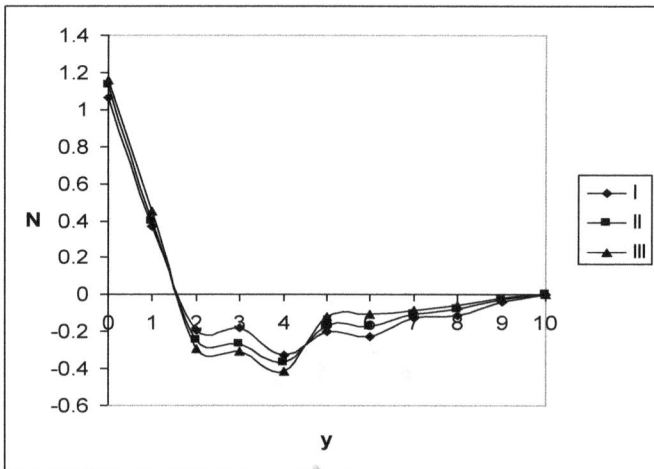

Fig. 8 : Variation of N with M

24

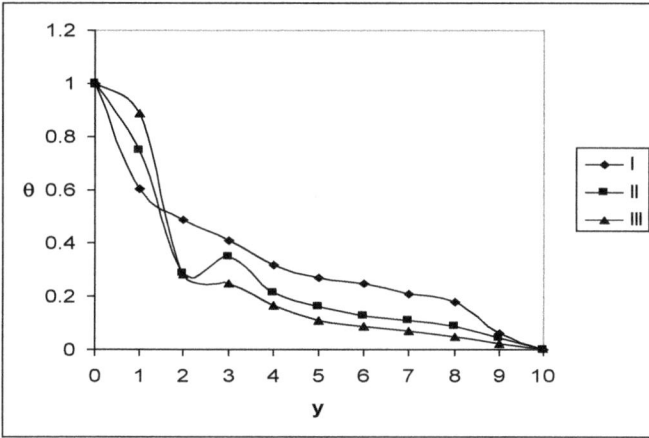

Fig. 9 : Variation of θ with M

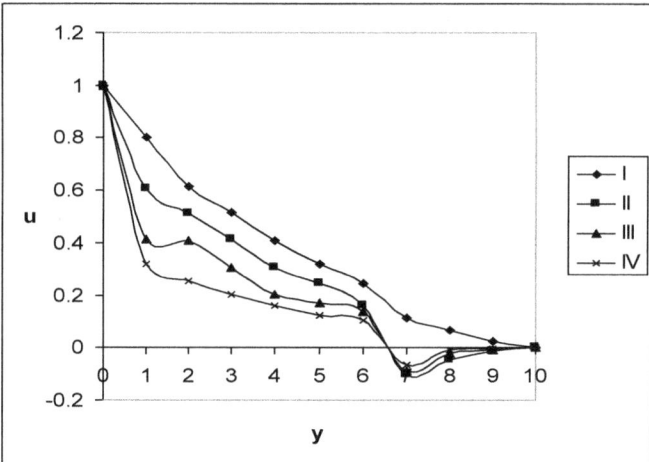

Fig. 10 : Variation of u with S

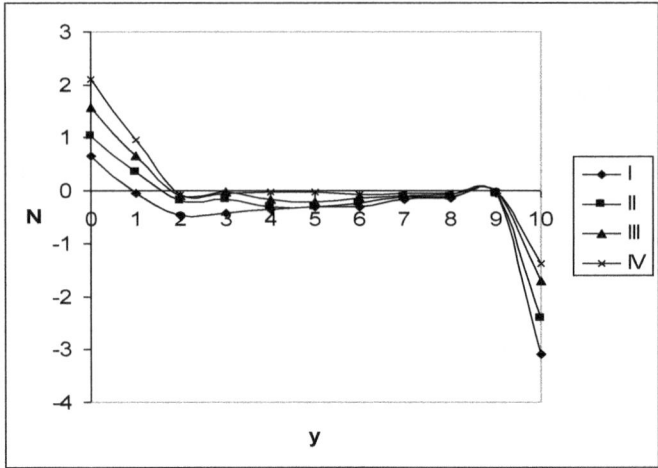

Fig. 11 : Variation of N with S

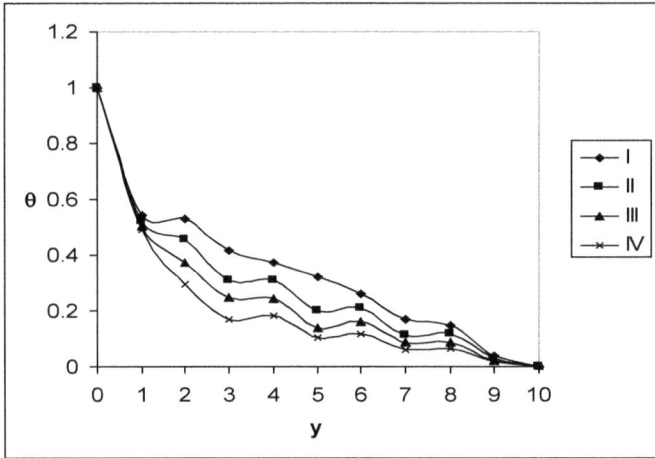

Fig. 12 : Variation of θ with S

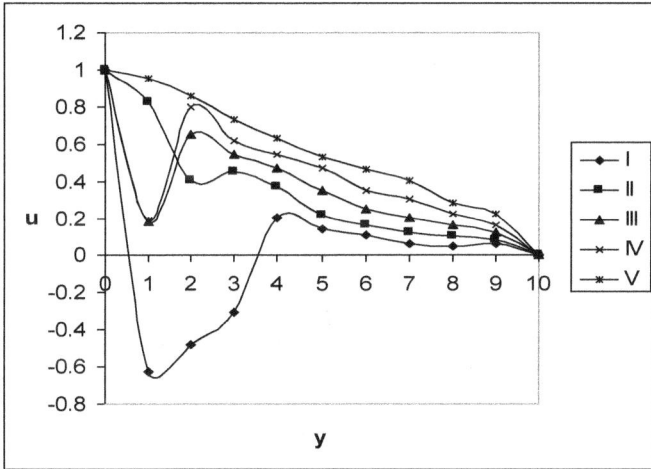

Fig. 13 : Variation of u with λ

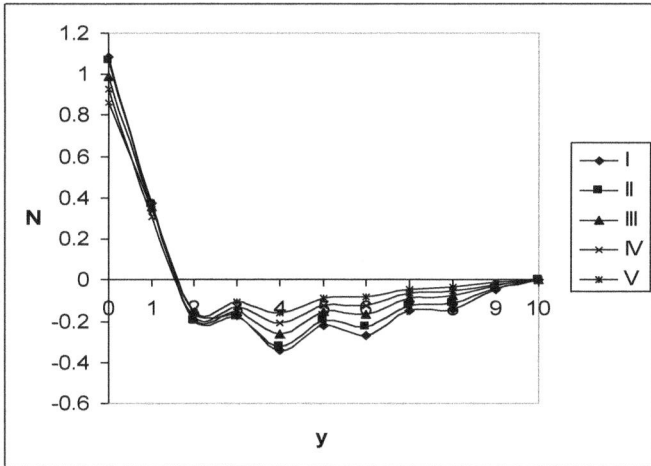

Fig. 14 : Variation of N with λ

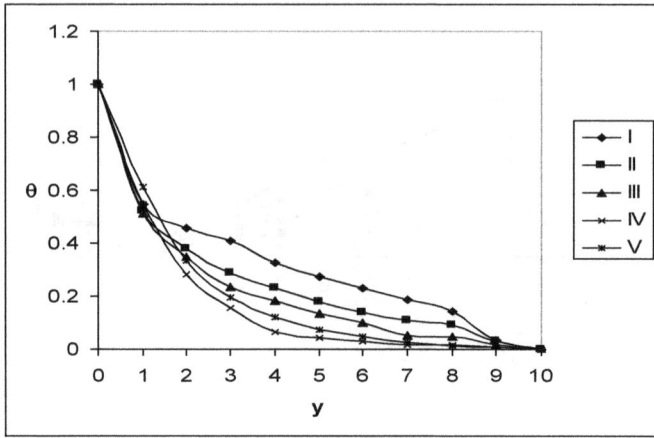

Fig.15 : Variation of θ with λ

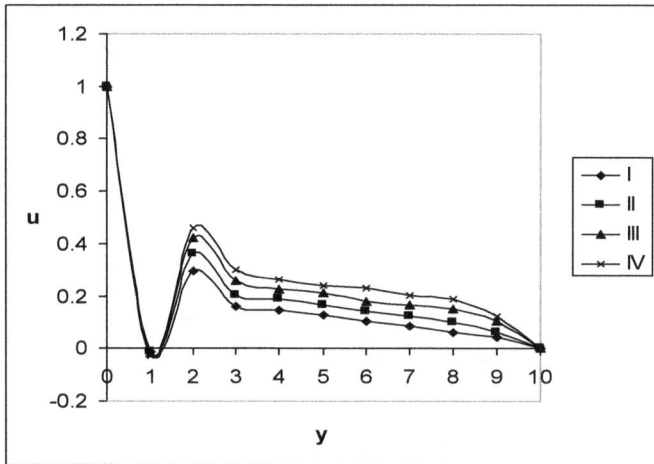

Fig. 16 : Variation of u with α

28

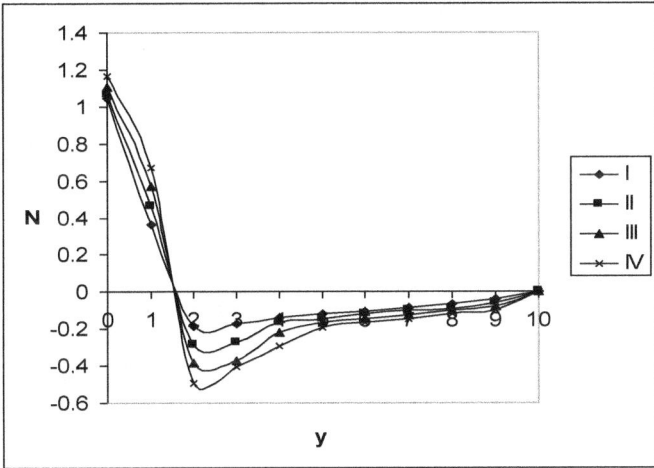

Fig. 17 : Variation of N with α

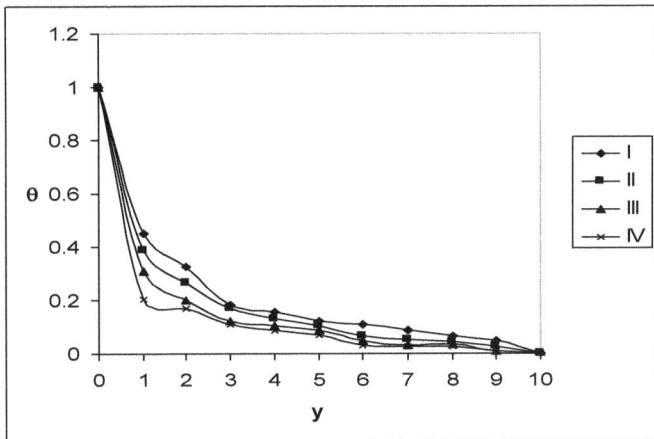

Fig. 18 : Variation of θ with α

The Velocity, Micro rotation and temperature functions as obtained by using the finite element method are shown in figs 1-18. The values of material parameter c1, c2 and c3 are taken to be fixed as 0.5, 2 and 0.05 respectively. The Prandtl number P_r and material parameter α, each are kept fixed at 1, whereas the effect of other important parameters, namely Grashof number G, Hartman number M, suction parameter λ, heat source parameter α, and radiation parameter N over these functions has been studied.

Fig.1 demonstrates the variation of velocity distribution with buoyancy parameter G. It is found that the velocity increase continuously with increasing G. This revolves that the connective parameter G increases the boundary layer thickness. Also near the boundary it is positive while away from the boundary it is negative i.e., it retards the flow field away from the boundary for small G. for small values of G $\leq$ 5 the velocity continuously decreases while for large values of G, it increases till it attains maximum value near the boundary and then decreases.

Fig.2 represents the micro rotation distribution with G. that is found that for all values of G the micro rotation positive in the vicinity of the boundary and negative for away from the boundary and finally goes to zero. An increasing Grashof number G reduces the micro rotation in the presence of magnetic field. Thus convection effects produce a reverse rotation away from the boundary.

Fig.3 represents the temperature distribution. The temperature continuously decreases. It has been found that the temperature depreciation increases in the Grashof number G. i.e., Temperature can be controlled by the convection parameter G. Thus we conclude that a desired temperature can be generated by controlling the buoyancy parameter G.

The variation of u, N and θ with Darcy parameter D^{-1} is shown in fig. 4 to 6. From fig. 4 we note that the velocity is negative in the vicinity of the boundary and it remains positive in the whole domain lesser the permeability of the porous medium.

30

Larger the velocity in the vicinity of the boundary and smaller the velocity in the whole domain.

The micro rotation is represented the fig. 5. It is observed that for any D^{-1} near the boundary micro rotation positive whereas away from the boundary. It becomes negative and finally goes to zero. Lesser the permeable porous medium larger the micro rotation the presence of the porous medium produce a reverse rotation for away from the boundary fig. 6 illustrates the temperature distribution on with different D^{-1}. It is observed that lesser the permeable porous medium larger the actual temperature near the boundary and smaller the actual temperature for away from the boundary. The variation of magnetic field on u, N, θ is shown in fig. 7, 8, 9. It is found that the velocity is −ve in the region 1 to 4 and positive for away from the boundary for $M \leq 5$ and for higher $M \geq 10$. It is negative in the whole domain. The reverse of the flow which appears in the region $1 \leq \eta$ 4 enlarges in its size also |u| experiences and enhancement in the region $1 \leq \eta \leq 4$ and depreciates the marginally for away from the boundary.

Fig.8 illustrates the micro rotation with Hartman number M. It is found that the micro rotation is positive in the vicinity of the boundary and negative in the whole domain. Thus the Lorentz force reduces a reverse rotation for away from the boundary. Also higher the Lorentz force larger the micro rotation in the region $1 \leq N \leq 4$ and smaller the magnetic field of micro rotation in the remaining region. Fig.9 illustrates the temperature distribution for a different M. It is found that the temperature positive in the whole domain. Higher the Lorentz force an increasing the Hartman, Number – M enhance the actual temperature in the vicinity of the boundary and depreciates the boundary in the whole domain.

Figs. 10-12 illustrate the variation of velocity micro rotation and temperature functions with the surface temperatures. Fig. 10 shows the variation of velocity distribution with S. It is clear that the velocity decreases with increasing parameters near the boundary it remains positive whereas away from it becomes negative and thus retards the flow. The velocity values corresponding to no-spin condition are the maximum.

Fig.11 represents the micro rotation distribution with S. It is found that the micro rotation is negative for S ≤ 0.5. In the neighborhood of the boundary and negative in the remaining region and increasing the surface parameters enhances the micro rotation near by the boundary and depreciates in the remaining region.

The temperature distribution is shown in fig. 12 for a different. It is found that an increasing the surface parameter S results in a depreciation in the actual temperature.

Figs. 13-15 represented the variation u, N and θ for different values of the section parameterλ. It shows that an injection increases a velocity increases while a reversed effect is observed in the case of suction. Large suction represented a reverse flow in the region 1 ≤ η ≤ 3. Thus the speed of flow can be controlled by suction, which is important for many engineering applications.

Fig.14 depicts the micro rotation distribution, with differentλ. It is found that a micro rotation is positive near by the boundary and negative in the remaining region. An injection increases the micro rotation experiences and enhances in the vicinity of the boundary and depreciates in the remaining region. An injection increases the micro rotation experiences and enhances in the vicinity of the boundary and depreciates in the remaining region, while the reverse is true for suction. It is observed that the micro rotation effects are more dominant in the vicinity of the boundary.

The micro rotation remains unaffected for large injection. Because for large injection micro effects distribution, its values continuously increase with increase in suction while a reverse pattern is noticed for suction. Thus a higher temperature can be obtained by using injection. Therefore we can conclude that suction and injection can be used for controlling/increasing the temperature function, which is required in many engineering applications. Like nuclear reactors, generators etc.,

The variation of u, N and θ with heat source parameter α is shown in fig 16-18 respectively. Fig. 16 represents the velocity distribution with heat source parameterα. It is found that the velocity exhibits a reversal flow in the vicinity of $\eta = 0$ and the region of reversal flow enlarges with increase in α. An increase in the strength of the heat generating source results in an enhancement in the velocity field in the whole domain.

Fig.17 illustrates micro rotation withα. It is found that micro rotation positive near by the boundary and negative in the remaining region. An increase in the strength of the heat source produces reverse rotation away from the boundary and this reverse rotation enhance with increase in the strength of heat generating source.

The temperature distribution with α is exhibited in fig. 18. It is found that the actual temperature experiences depreciation with increase inα. Thus the presence of the heat source in the fluid region reduces actual temperature.

The skin friction at the plate y = 0 is shown in Table 1 and 2 for different values of G, D^{-1}, M, λ andα. It is found that the skin friction at y = 0 is negative for all variations. It is found that skin friction enhances with increases. In G > 0 and reduces with G < 0. The variation of λ with D^{-1} shows that in the heating case the skin friction reduces with D^{-1} and in the cooling case λ enhances with $D^{-1} \leq 3 \times 10^2$ and depreciates with higher $D^{-1} \geq 5 \times 10^2$. The variation of λ with Hartman number M shows that higher the Lorentz force smaller λ and larger λ with respect to variation of λ with suction parameterλ. We find that the skin friction numerically increases with injection while it depreciates and enhances with suction parameter. An increasing in the heat source parameter α and enhance it in the cooling of the plate. The Nusselt

33

number with represent the heat transfer $y = 0$ is evaluated for different parameter values as shown in fig. 3 and 4. It is found that the rate of heat transfer enhances with $G < 0$ and reduces with $G > 0$. An increasing lesser the permeable porous medium are higher Lorentz force larger $|\lambda|$ for $G > 0$ and enhances depreciates with $G < 0$. Thus the skin friction at the plate reduces with suction parameter and enhances with injection parameter and the drag reduces effectively with convection parameter as well as the suction parameter as well as with injection. Also the skin friction depreciates the parameter α. Thus the drag reduces effectively with convection parameter, Darcy parameter, and Hartmann number.

Thus the drag can be reduce effectively with convection parameter $G > 0$. While it enhances with $G < 0$, D^{-1} and M in Table – 3 and 4. The couple stress at $y = 0$ is shown in tables 5 and 6 for different value of G, D^{-1}, M, λ, S and α. It is found that the couple stress depreciates with $G > 0$ and enhance with $G < 0$. A variation of couple stress with D^{-1} and M shows that lesser the permeability of porous medium of higher the Lorentz force larger the couple stress for $G > 0$ and smaller for $G < 0$ an increasing the parameter S depreciates the couple stress for all G. the couple stress numerically reduces for with an increases with increases heat source parameter α in both heating and cooling case. Also the couple stress reduces with suction parameter where as it enhances increase in suction table 5 and 6.

REFERENCES :

[1] ARIMAN, T. TURK, M.A. SYLVESTER, N.D. Review article – applications of micro continuous fluid mechanics, Int. J. Eng. sci. 12 (1974) 273.

[2] ABO – ELDAHAB, E.M. and CHONAIM, A.F : convective heat transfer in an electrically conducting micro polar fluid at a stretching surface with uniform free stream. Appl. Math comput. 137, (2003). 323-336.

[3] AGARWAL, R.S. BHARGAVA, R. BALAJI, A.V.S. finite element solution of flow and heat transfer of micro polar fluid over a stretching sheet. Int. J. Eng. Sci. 27 (1989) 1421.

[4] BHARGAVA, R. KUMAR, L. TAKHAR, H.S. Numerical solution of free convection MHD micro polar fluid flow between two parallel porous vertical plates. Int. J. Eng. sci. 41 (2003) 123.

[5] DESSEAUX, A. and KELSON, N.A: flow of a micro polar fluid bounded by a stretching sheet. Anziam J. 42CE. C536 – C560.

[6] ERINGEN, A.C. simple micro fluids, Int. J. Eng. sci. 2 (1964) 205.

[7] ERINGEN, A.C. theory of micro polar fluids J. Math. Mech. 16 (1966).

[8] ERINGEN, A.C. theory of thermo-micro fluids, J. Math. Anal. Apple. 38 (1972) 480.

[9] FANSLER, K.S. DAMBERG, J.E. non similar moving wall boundary layer problem, quart, Appl. Math. 34 (1976) 305.

[10] GORLA, R.S.R. GHORASHI, B. WANGSKARN, P. mined convection in vertical internal flow of a micro polar fluid, Int. J. Eng. sci. 27 (1989) 1553.

[11] GILL, W.N. CASAL, E.D. a theoretical investigation of natural convection effects in forced horizontal flows. AlchE J. 8 (1962) 513.

[12] HADY, F.M : on the solution of heat transfer to micro polar fluid from a non-isothermal stretching sheet with injection. International journal of numerical methods for heat and fluid flow 6(6). (1966). 99-104.

[13] HOYT, J.W. FABULA, A.G. the effect of additive on flid friction, us Naval ordnance test station Report, 1964.

[14] KELSON, N.A. DESSEAUN, A. effect of surface conditions on flow of a micro polar fluid driven by a porous stretching sheet, Int. J. Eng. sci. 39 (2001) 1881.

[15] MOHAMADEIN, A.A. and GORLA, R.S.A. Effects of transverse magnetic field on mined convection in a micro polar fluid on mined convection in a micro polar fluid on a horizontal plate with vectored mass transfer. Acta mechanical,118 (1996). 1-12, [19] Siddeshwar, P.G. and Pranesh, S: Magneto convection in a micro polar fluid. Int. J. Eng. sci. 36 (10), (1998) 1173-1181.

[16] MOHAMMADEIN, A.A. and GORLA, R: Heat transfer in a micro polar fluid over a stretching sheet with viscous dissipation and internal heat generation. International journal of numerical methods for Heat and Fluid flow, 11(1) (2001), 50-58.

[17] NA, T.Y. and POP, I: Boundary-layer flow of micro polar fluid due to a stretching wall. Archives of Applied mechanics, 67 (4), (1977). 229-236.

[18] POWER, H. micro polar fluid model for the brain fluid dynamics, Int. conf. on bio-fluid mechanics, 1998, 04.

[19] YUCEL, A. mined convection in micro polar fluid flow over a horizontal plate with surface mass transfer, Int. J. Eng. sci. 27 (1989) 1593.

CHAPTER – II

MICRO CONVECTIVE HEAT TRANSFER FLOW OF AN ELECTRICALLY CONDUCTING MICROPOLAR FLUID THROUGH A POROUS MEDIUM IN A VERTICAL CHANNEL WITH HEAT SOURCES – A FINITE ELEMENT STUDY

2.1. INTRODUCTION:

In the recent years, the study of free convection phenomenon has been the object of extensive research. The intensity of research in this field is due to enhanced concerns in science and technology about buoyancy-induced motions in the atmosphere, in bodies of water and in quasi-solid bodies such as earth. Heat transfer effects under the conditions of free convection are now dominant in many engineering applications such as rocket nozzles, cooling of nuclear reactors, high speed aircrafts and their atmospheric reentry, high sinks in turbine blades, chemical devices and process equipment, formation and dispersion of fog, distribution of temperature and moisture over agricultural fields and graves of fruit trees, damage of crops due to freezing and pollution of the environment and so on. Hence a thorough investigation and knowledge of heat transfer process must be acquired in order to be able to design heat exchangers, bearing etc., so that no overheating or damage is caused to the components.

Simple problems on the flow of such fluids were studies by a number of researchers and a review of this work was given by Ariman et. al.,[1].

Balaram and Sastry [3] studied the fully developed free convection flow in a micro polar flow. Later Jena and Mathur [6] examined the free convective heat transfer to a micro polar fluid along a non-isothermal vertical plate. Agarwal and Dhanapal [2] have analized the effect of temperature dependent heat sources on the fully developed free convection micro polar fluid flow when a constant suction or injection is applied on the plates and the fluid.

The extension of above type of flows to include magneto hydrodynamic effects has become important due to several engineering applications such as in MHD generators, designing cooling system for nuclear reactors, flow meters, etc., where the micro concentration provides an important in such equipment's. Several investigators have made theoretical and experimental studies of micro polar flow in the presence of a transverse magnetic field during the last three decades.

Kasivishwanathan and Gandhi [7] have studies a class of exact solutions for the MHD flow of a micro polar fluid confined between two infinite, insulated, parallel, non-coaxially rotating disks. Mohammadein and Gorla [8] investigated the transverse magnetic field on mined convection in a micro polar fluid flowing on a horizontal plate with vectored mass transfer. Later Gorla and Takhar [5] examined the simultaneous occurrence of buoyancy and magnetic forces in the flow of an electrically conduction micro polar fluid along a hot vertical plate in the presence of a strong crass magnetic field. The results indicate that the micro polar fluids reduce drang and surface heat transfer rate. The general theory of magneto-micro polar fluids can be found in Erigen [4], convective heat transfer in a micro polar fluid flowing in an annulus of convection in the laminar boundary layer flow of a thermo micro polar fluid past a non-uniformly heated vertical flat plate. Rama Bhargava et. al., [9] have obtained numerical solution free convection MHD micro polar flow between two parallel porous vertical plates.

In this chapter we consider the fully developed electrically conducting micro fluid flow through a porous medium confined in a vertical channel bounded by two porous parallel plates is studied in the presence of heat generating sources including the effect of frictional heating and the influence of uniform transverse magnetic field. The transport equations are momentum, micro rotation and energy are solved by employing galerkin finite element analysis with quadratic approximation functions, profiles for velocity, micro rotation and temperature and presented for different values of Hartman number H. micro polar parameter R, Darcy parameter D^{-1} heat source parameter γ. The skin fiction Nusselt number and couple stress at the plates are numerically evaluated for different sets of the parameter.

2. 2. FORMULATION OF THE PROBLEM:

Consider the fully developed steady, laminar free convection flow of an incompressible micro polar fluid through a porous medium flowing between two infinite parallel porous flat plates distance h a part and oriented in the direction of the body force. The plates are maintained at constant temperature T_1 and T_2 in the presence of strong magnetic field H_0 normal to the plate. The x axis is taken along one of the plates and y axis normal to it. Since the boundaries in the x direction are of infinite dimensions therefore without any loss of generality we assume that the physical quantities, for example velocity, micro rotation and temperature depends on y only. The velocity field is taken to be (u, v, 0) and micro rotation as (0, 0, N).

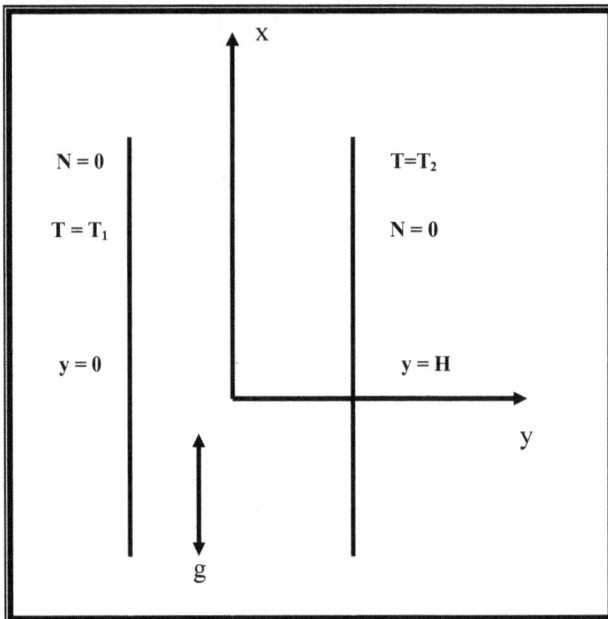

It is assumed that the fluid possess constant properties except the density variation due to temperature difference is used only to express the body force term as the buoyancy term. Thus the governing equations of such type of flow can be written as:

$$\frac{\partial v}{\partial y} = 0, \text{giving } v = v_0 \text{ (constant)} \qquad 2.1$$

$$\rho v_0 \frac{\partial u}{\partial y} = (\mu + k)\frac{\partial^2 u}{\partial y^2} + \frac{\partial N}{\partial y} + \rho f - \frac{\partial p}{\partial x} + \sigma H_0^2 u - \left(\frac{\mu}{k_1}\right) u \qquad 2.2$$

$$\rho j v_0 \frac{\partial N}{\partial y} = \gamma \frac{\partial^2 N}{\partial y^2} - k\left(2N + \frac{\partial u}{\partial y}\right) \qquad 2.3$$

$$\rho C_p v_0 \frac{\partial T}{\partial y} = k_f \frac{\partial^2 T}{\partial y^2} + \left[(\mu + k)\left(\frac{\partial u}{\partial y}\right)^2 + r\left(\frac{\partial N}{\partial y}\right)^2 + 2k\left(N^2 + N\frac{\partial u}{\partial y}\right)\right] + \gamma_0 v_0 (T - T_0) \quad 2.4$$

where γ_0 is the constant of proportionality, $\gamma_0 v_0 (T - T_0)$. The amount of heat generated per unit volume in unit time is assumed to be a linear function of temperature, p the pressure, ρ the density, f the body force, C_p the specific heal, μ the coefficient of viscosity, k the gyro viscosity γ the material constant, j the micro inertia and k_f the coefficient of thermal conductivity and T_0 is the temperature in hydrostatic state.

The boundary conditions are given by

at $y = 0$, $u = 0$, $v = v_0$, $T = T_1$, $N = 0$

at $y = h$, $u = 0$, $v = v_0$, $T = T_2$, $N = 0$ \qquad 2.5

where $v_0 < 0$ implies injection at $y = h$ and suction at $y = 0$ while $v_0 > 0$ represents injection at $y = 0$ and suction at $y = h$.

The body force term is expressed as a buoyancy term.

41

$$\rho f - \frac{\partial p}{\partial x} = \rho \beta (T - T_0) f_x \qquad\qquad 2.6$$

Where $f_x = -f$, $PD = (P - P_0)$ is the pressure difference, P_0 being the hydrostatic pressure and β is the coefficient of volumetric expansion.

As the plates are infinitely long, the pressure p can be taken equal to the hydro static pressure p_0, i.e. $p_0 = 0$.

Using equation (3.6), equations (3.2) – (3.4) can be written as

$$\frac{\partial^2 u}{\partial y^2} - \frac{\rho v_0}{(\mu + k)} \frac{du}{dy} + \frac{k}{\mu + k} \frac{dN}{dy} + \frac{\rho \beta f_x (T - T_0)}{\mu + k} - \frac{\sigma H_0^2 u}{\mu + k} - \left(\frac{\mu}{k_1 (\mu + k)} \right) u = 0 \qquad 2.7$$

$$\frac{d^2 N}{dy^2} - \frac{\rho j v_0}{\gamma} \frac{dN}{dy} - \frac{k}{\gamma} \left(2N + \frac{du}{dy} \right) = 0 \qquad\qquad 2.8$$

$$\frac{d^2 (T - T_0)}{dy^2} - \frac{\rho C_p v_0}{k_f} \frac{d(T - T_0)}{dy} + \left[\frac{(\mu + k)}{k_f} \left(\frac{\partial}{\partial y} \right)^2 + \frac{\gamma}{k_f} \left(\frac{\partial N}{\partial y} \right)^2 + \frac{2k}{k_f} \left(N^2 + N \frac{\partial u}{\partial y} \right) \right]$$
$$+ \frac{r_0 v_0 (T - T_0)}{k_f} = 0 \qquad\qquad 2.9$$

Introducing the following dimensionless variables

$$y = \eta h, \qquad\qquad u = \frac{k_f U}{\rho f_x \beta h^2},$$

$$\qquad\qquad\qquad\qquad\qquad\qquad\qquad\qquad 2.10$$

$$(T - T_0) = \frac{k_f \mu T^*}{\rho^2 f_x^2 \beta^2 h^4}, \qquad N = \frac{k_f N^*}{\rho f_x \beta h^3}$$

and substituting the above transformations into the equations (2.7) – (2.9), we obtain

$$\frac{d^2 u}{d\eta^2} - \frac{Re}{1 + R} \frac{du}{d\eta} + \frac{T^*}{1 + R} + \frac{R}{1 + R} \frac{dN^*}{d\eta} - C(H_a^2 + D^{-1}) v = 0 \qquad 2.11$$

$$\frac{d^2 N}{d\eta^2} - \frac{Re \beta}{A} \frac{dN^*}{d\eta} - \frac{R}{A} \left(2N^* + \frac{du}{d\eta} \right) = 0 \qquad\qquad 2.12$$

$$\frac{d^2T^*}{d\eta^2} - \mathrm{Re}\,\mathrm{Pr}\,\frac{dT^*}{d\eta} + (1+R)\left(\frac{du}{d\eta}\right)^2 + A\left(\frac{dN^*}{d\eta}\right)^2$$

$$+ 2R\left(N^* + N^*\frac{du}{d\eta}\right) + \gamma_1\,\mathrm{Re}\,\mathrm{Pr}\,T^* = 0$$

2.13

where $R = \left(\dfrac{k}{\mu}\right)$ is the micro polar parameter, $\mathrm{Re} = \left(\dfrac{\rho v_0 h}{\mu}\right)$ The cross flow Reynolds

number, $A = \left(\dfrac{\gamma}{\mu h^2}\right)$, $B = \left(\dfrac{j}{h^2}\right)$ and $C = \left(\dfrac{h^2}{1+R}\right)$ The micro polar material constants,

$\mathrm{Pr} = \dfrac{\mu C_p}{k_f}$ The Prandtl number, $\gamma_1 = \left(\dfrac{\gamma_0 h}{\rho C_p}\right)$ The dimensionless vertical distance,

$H_a = \left(\dfrac{\sigma H_0^2}{\mu}\right)$ is the local Hartmann number, $D^{-1} = \dfrac{h^2}{k_1}$, k_1 = permeability of the porous

medium.

The corresponding boundary conditions given in equation (2.5) then reduces to

at $\qquad \eta = 0, \qquad U = 0, \qquad N^* = 0 \qquad\qquad T^* = Q$

$\qquad\qquad \eta = 1, \qquad U = 0, \qquad N^* = 0 \qquad\qquad T^* = \varepsilon\,Q$

where $Q = \left(\dfrac{\mathrm{Pr}\,\mathrm{Gr}\,\beta\,f_x\,h}{c_p}\right)$ is a dimensionless group, $\mathrm{Gr} = \left(\dfrac{\rho^2\beta\,h^2\Delta T}{\mu_2}\right)$ is the Grashof

number and $\in = \left(\dfrac{T_2 - T_0}{T_1 - T_0}\right)$ is the non-dimensional heating parameter.

2.3. FINITE ELEMENT ANALYSIS OF THE PROBLEM:

To solve these differential equations with the corresponding boundary conditions, we assume if u^i, N^i, θ^i are the approximations of u, N and θ we define the errors (residual) $E_1^i, E_2^i, E_3^i, E_4^i$ as

$$E_u^i = \frac{d}{d\eta}\left(\frac{du^i}{d\eta}\right) - \frac{\mathrm{Re}}{1+R}\frac{du^i}{d\eta} + \frac{\theta i}{1+R} + \frac{R}{1+R}\frac{dN^2}{d\eta} - C\,(H_a^i + D^{-1})u = 0 \qquad 3.1$$

43

$$E_N^i = \frac{d}{d\eta}\left(\frac{d N^i}{d\eta}\right) - \frac{Re\,\beta}{A}\frac{d N^i}{d\eta} - \frac{R}{A}\left(2N^i + \frac{d u^i}{d\eta}\right) = 0 \qquad\qquad 3.2$$

$$E_\theta^i = \frac{d}{d\eta}\left(\frac{d\theta^i}{d\eta}\right) - Re\,Pr\,\frac{d\theta^i}{d\eta} + (1+R)\left(\frac{d u^i}{d\eta}\right) + A\left(\frac{d N^i}{d\eta}\right)^2 + 2R\left(N^{i^2} + N^i\frac{d u^i}{d\eta}\right)$$
$$+\,\gamma_1\,Re\,Pr\,\theta^i = 0 \qquad\qquad 3.3$$

where

$$\left.\begin{array}{l} u^i = \sum\limits_{k=1}^{3} u_k \psi_k \\[4mm] N^i = \sum\limits_{k=1}^{3} N_k \psi_k \\[4mm] \theta^i = \sum\limits_{k=1}^{3} \theta_k \psi_k \end{array}\right\} \qquad\qquad 3.4$$

These errors are orthogonal to the weight function over the domain of e^1 under Galerkin finite element technique we choose the approximation functions as the weight function multiply both sides of the equations (3.1 – 3.3) by the weight function i.e. each of the approximation function ψ_j^i and integrate over the typical two nodded linear element (η_e, η_{e+1}) we obtain

$$\int\limits_{\eta_e}^{\eta_{e+1}} E_u^i\,\psi_j^i\,dy = 0 \qquad\qquad (i = 1,2,3,4) \qquad\qquad 3.5$$

$$\int\limits_{\eta_e}^{\eta_{e+1}} E_N^i\,\psi_j^i\,dy = 0 \qquad\qquad (i = 1,2,3,4) \qquad\qquad 3.6$$

$$\int\limits_{\eta_e}^{\eta_{e+1}} E_\theta^i\,\psi_j^i\,dy = 0 \qquad\qquad (i = 1,2,3,4) \qquad\qquad 3.7$$

where

$$\int\limits_{\eta_e}^{\eta_{e+1}}\left[\frac{d}{d\eta}\left(\frac{d u^i}{d\eta}\right) - \frac{Re}{1+R}\frac{d u^i}{d\eta} + \frac{\theta^i}{1+R} + \frac{R}{1+R}\frac{d N^i}{d\eta} - C\left(H_a^2 + D^2\right)u\right]\psi_j^i\,d\eta = 0 \quad 3.8$$

$$\int\limits_{\eta_e}^{\eta_{e+1}}\left[\frac{d}{d\eta}\left(\frac{d N^i}{d\eta}\right) - \frac{Re\,\beta}{A}\frac{d N^i}{d\eta} - \frac{R}{A}\left(2N^i + \frac{d u^i}{d\eta}\right)\right]\psi_j^i\,d\eta = 0 \qquad\qquad 3.9$$

$$\int_{\eta_e}^{\eta_{e+1}} \left[\frac{d}{d\eta}\left(\frac{d\theta^i}{d\eta}\right) - \text{Re Pr} \frac{d\theta^i}{d\eta} + (1+R)\left(\frac{du^i}{d\eta}\right)^2 + A\left(\frac{dN^i}{d\eta}\right)^2 + 2R\left(N^{i^2} + N^i \frac{du^i}{d\eta}\right) + r_i \text{ Re Pr } \theta^i \right] \psi_j^i \, d\eta = 0 \quad 3.10$$

Following the Galerkin weighted residual method and integration by parts method to the equations (3.8) – (3.10) we obtain

$$\int_{\eta_e}^{\eta_{e+1}} \frac{d\Psi_j^i}{d\eta} \frac{d\psi^i}{d\eta} d\eta - \int_{\eta_e}^{\eta_{e+1}} \frac{\text{Re}}{1+R} \frac{du^i}{d\eta} \Psi_j^i d\eta + \int_{\eta_e}^{\eta_{e+1}} \frac{\theta^i \Psi_j^i}{1+R} d\eta + \int_{\eta_e}^{\eta_{e+1}} \frac{R}{1+R} \frac{dN^i}{d\eta} \Psi_j^i d\eta$$
$$- \int_{\eta_e}^{\eta_{e+1}} C(H_a^2 + D^{-1}) u^i \, \Psi_j^i \, d\eta = Q_{1,j} + Q_{2,j} \qquad 3.11$$

Where $\quad - Q_{1,j} = \Psi_j(\eta_e) \dfrac{du^i}{d\eta}(\eta_e)$

$$Q_{1,j} = \Psi_j(\eta_e) \frac{du^i}{d\eta}(\eta_e)$$

$$\int_{\eta_e}^{\eta_{e+1}} \frac{d\Psi_j^i}{d\eta}\left(\frac{dN^i}{d\eta}\right) d\eta - \int_{\eta_e}^{\eta_{e+1}} \frac{\text{Re } \beta}{A} \frac{dN^i}{d\eta} \Psi_j^i \, d\eta - \int_{\eta_e}^{\eta_{e+1}} \frac{R}{A}\left(2N^i \Psi_j^i + \frac{du^i}{d\eta}\right) d\eta = R_{1,j} + R_{2,j} \quad 3.12$$

where $\quad - R_{1,j} = \Psi_j(\eta_e) \dfrac{dN^i}{d\eta}(\eta_e) + \Psi_j(\eta_e) \dfrac{d\theta^i}{d\eta}(\eta_e)$

$$R_{2,j} = \Psi_j(\eta_{e+1}) \frac{dN^i}{d\eta}(\eta_{e+1}) + \Psi_j(\eta_{e+1}) \frac{d\theta^i}{dy}(\eta_{e+1})$$

$$\int_{\eta_e}^{\eta_{e+1}} \frac{d\Psi_j^i}{d\eta} \frac{d\theta^i}{d\eta} d\eta - \text{Re Pr} \int_{\eta_e}^{\eta_{e+1}} \frac{d\theta^i}{d\eta} \psi_j^i \, d\eta + \int_{\eta_e}^{\eta_{e+1}} \left(\frac{du^i}{d\eta}\right)^2 \psi_j^i d\eta + \int_{\eta_e}^{\eta_{e+1}} A\left(\frac{dN^i}{d\eta}\right)^2 \psi_j^i d\eta$$
$$+ \int_{\eta_e}^{\eta_{e+1}} 2R\left(N_i^2 + N^i \frac{du^i}{d\eta}\right) \psi_j^i d\eta + \int_{\eta_e}^{\eta_{e+1}} r_i \text{ Re Pr } \theta_k \psi_j^k \, d\eta = S_{1,j} + S_{2,j} \qquad 3.13$$

where $\quad - S_{1,j} = \Psi_j(\eta_e) \dfrac{d\theta^i}{d\eta}(\eta_e) + \Psi_j(\eta_e) \dfrac{dN^i}{d\eta}(\eta_e)$

$$S_{2,j} = \Psi_j(\eta_{e+1}) \frac{d\theta^i}{d\eta}(\eta_{e+1}) + \Psi_j(\eta_{e+1}) \frac{dN^i}{dy}(\eta_{e+1})$$

45

Making use of equations (3.4) we can write above equations as

$$\sum_{k=1}^{3} u_k \int_{\eta_e}^{\eta_{e+1}} \frac{d\psi_j^i}{d\eta} \frac{d\psi_k}{d\eta} d\eta - \sum_{k=1}^{3} \frac{Re\, u_k}{1+R} \int_{\eta_e}^{\eta_{e+1}} \frac{d\psi_k}{d\eta} \psi_j^i \, d\eta + \sum_{k=1}^{3} \theta_k \int_{\eta_e}^{\eta_{e+1}} \frac{\psi_j^i}{1+R} \psi_k \, d\eta + \sum_{k=1}^{3} \frac{R}{1+R} N_k$$

$$\int_{\eta_e}^{\eta_{e+1}} \frac{d\psi_k}{d\eta} \psi_j^i \, d\eta - \sum_{k=1}^{3} u_k \int_{\eta_e}^{\eta_{e+1}} C(H_a^2 + D^{-1}) u^i \psi_j^i \, d\eta = Q_{1,j} + Q_{2,j} \qquad 3.14$$

$$\sum_{k=1}^{3} N_k \int_{\eta_e}^{\eta_{e+1}} \frac{d\psi_j^i}{d\eta} \frac{d\psi_k}{d\eta} d\eta - \sum_{k=1}^{3} N_k \int_{\eta_e}^{\eta_{e+1}} \frac{Re\,\beta}{A} \frac{d\psi_k}{d\eta} \psi_j^i \, d\eta - \sum_{k=1}^{3} N_k \int_{\eta_e}^{\eta_{e+1}} \frac{R}{A} (2\psi_k \, \psi_j^i) \, d\eta$$

$$- \sum_{k=1}^{3} u_k \int_{\eta_e}^{\eta_{e+1}} \frac{d\Psi_k}{d\eta} \psi_j^i \, d\eta = R_{1,j} + R_{2,j} \qquad 3.15$$

$$\sum_{k=1}^{3} \theta_k \int_{\eta_e}^{\eta_{e+1}} \frac{d\psi_j^i}{d\eta} \frac{d\psi_k}{d\eta} d\eta - \sum_{k=1}^{3} \theta_k \int_{\eta_e}^{\eta_{e+1}} Re\, Pr \frac{d\psi_k}{d\eta} \psi_j^i \, d\eta + \sum_{k=1}^{3} u_k^2 \int_{\eta_e}^{\eta_{e+1}} \left(\frac{d\psi_k}{d\eta}\right)^2 \psi_j^i \, d\eta$$

$$- \sum_{k=1}^{3} N_k^2 \int_{\eta_e}^{\eta_{e+1}} A \left(\frac{d\psi_k}{d\eta}\right)^2 \psi_j^i \, d\eta + \sum_{k=1}^{3} N_k^2 \int_{\eta_e}^{\eta_{e+1}} 2R\, \psi_k^2 \, \psi_j^i \, d\eta + \sum_{k=1}^{3} N_k u \int_{\eta_e}^{\eta_{e+1}} 2R \frac{d\psi_k}{d\eta} \psi_j^i \, d\eta \quad 3.16$$

$$+ \sum_{k=1}^{3} \theta_k \int_{\eta_e}^{\eta_{e+1}} r_1 \, Re\, Pr\, \psi_k \, \psi_j^i \, d\eta = S_{1,j} + S_{2,j}$$

Choosing different Ψ_j^i's corresponding to each element η_e in the equation (3.14) yields a local stiffness matrix of order 3×3 in the form

$$(f_{i,j}^k)(u_i^k) - \frac{R}{1+R} g_{i,j}^k (u_i^k - N_i^i - R\theta_i^k) - C(H_a^2 + D^{-1})(m_{i,j}^k)(u_i^k) = Q_{1,j}^k + Q_{2,j}^k \qquad 3.17$$

Likewise the equation (3.15) & (3.16) gives rise to stiffness matrices

$$(e_{ij}^k)(N_i^k) - \frac{Re\,\beta}{A} Du(t_{ij}^k)(u_i^k) = R_{1J}^k + R_{2J}^k \qquad 3.18$$

$$(l_{ij}^k)(\theta_i^k) - Re\, Pr\,(t_{ij}^k)(\theta_i^k) + (1+R)(\theta_i^k)^2 + AD^2(N_i^k)^2 + 2R(N_i^{k^2} + N_i^k\, Du_i^k) = S_{1,J}^k + S_{2,J}^k \quad 3.19$$

where

$\left(f_{i,J}^k\right), \left(g_{i,J}^k\right), \left(m_{i,J}^k\right), \left(n_{i,J}^k\right), \left(e_{i,J}^k\right), \left(t_{iJ}^k\right)$ are 3×3 matrices and $\left(Q_{2,J}^k\right), \left(Q_{1,J}^k\right), \left(R_{2,J}^k\right), \left(R_{1J}^k\right), \left(S_{2J}^k\right)$ and $\left(S_{1J}^k\right)$ are 3×1 column matrices and such stiffness matrices (3.17) – (3.19) in terms of local nodes in each element are assembled using

46

inter element continuity and equilibrium conditions to obtain the coupled global matrices in terms of the global nodal values of k, θ & φ. In case we choose n-quadratic elements then the global matrices are of order 2n+1. The ultimate coupled global matrices are solved to determine the unknown global nodal values of the velocity, temperature and concentration in fluid region. In solving these global matrices an iteration procedure has been adopted to include the boundary and effects in the porous region.

The shape functions corresponding to

$$\Psi_1^1 = \frac{(y-4)(y-8)}{32}$$

$$\Psi_2^1 = \frac{(y-12)(y-16)}{32}$$

$$\Psi_3^1 = \frac{(y-20)(y-24)}{32}$$

$$\Psi_1^2 = \frac{(y-2)(y-4)}{8}$$

$$\Psi_2^2 = \frac{(y-6)(y-8)}{8}$$

$$\Psi_2^3 = \frac{(y-10)(y-12)}{8}$$

$$\Psi_1^3 = \frac{(3y-4)(3y-8)}{32}$$

$$\Psi_2^3 = \frac{(3y-12)(3y-16)}{32}$$

$$\Psi_3^3 = \frac{(3y-20)(3y-24)}{32}$$

$$\Psi_1^4 = \frac{(y-1)(y-2)}{2}$$

$$\Psi_2^4 = \frac{(y-3)(y-4)}{2}$$

$$\Psi_3^4 = \frac{(y-5)(y-6)}{2}$$

$$\Psi_1^5 = \frac{(5y-4)(5y-8)}{32}$$

$$\Psi_2^5 = \frac{(5y-12)(5y-16)}{32}$$

$$\Psi_3^5 = \frac{(5y-20)(5y-24)}{32}$$

2.4. STIFFNESS MATRICES:

The global matrix for θ is

$$A_3 X_3 = B_3$$

4.1

47

The global matrix for is

$$A_4 X_4 = B_4 \qquad\qquad\qquad 4.2$$

The global matrix h is

$$A_5 X_5 = B_5 \qquad\qquad\qquad 4.3$$

where

$$
A_3 = \begin{pmatrix}
-1 & a_{12} & a_{13} & 0 & 0 & 0 & 0 & 0 & 0 & 0 & 0 \\
0 & a_{22} & a_{23} & 0 & 0 & 0 & 0 & 0 & 0 & 0 & 0 \\
0 & a_{32} & a_{33} & a_{34} & a_{35} & 0 & 0 & 0 & 0 & 0 & 0 \\
0 & 0 & a_{43} & a_{44} & a_{45} & 0 & 0 & 0 & 0 & 0 & 0 \\
0 & 0 & a_{53} & a_{54} & a_{55} & a_{56} & a_{57} & 0 & 0 & 0 & 0 \\
0 & 0 & 0 & 0 & a_{65} & a_{66} & a_{67} & 0 & 0 & 0 & 0 \\
0 & 0 & 0 & 0 & a_{75} & a_{76} & a_{77} & a_{78} & a_{79} & 0 & 0 \\
0 & 0 & 0 & 0 & 0 & 0 & a_{87} & a_{88} & a_{89} & 0 & 0 \\
0 & 0 & 0 & 0 & 0 & 0 & a_{97} & a_{98} & a_{99} & a_{910} & 0 \\
0 & 0 & 0 & 0 & 0 & 0 & 0 & 0 & a_{109} & a_{1010} & 0 \\
0 & 0 & 0 & 0 & 0 & 0 & 0 & 0 & a_{119} & a_{1110} & -1
\end{pmatrix}
$$

$$
A_4 = \begin{pmatrix}
-1 & b_{12} & b_{13} & 0 & 0 & 0 & 0 & 0 & 0 & 0 & 0 \\
0 & b_{22} & b_{23} & 0 & 0 & 0 & 0 & 0 & 0 & 0 & 0 \\
0 & b_{32} & b_{33} & b_{34} & b_{35} & 0 & 0 & 0 & 0 & 0 & 0 \\
0 & 0 & b_{43} & b_{44} & b_{45} & 0 & 0 & 0 & 0 & 0 & 0 \\
0 & 0 & b_{53} & b_{54} & b_{55} & b_{56} & b_{57} & 0 & 0 & 0 & 0 \\
0 & 0 & 0 & 0 & b_{65} & b_{66} & b_{67} & 0 & 0 & 0 & 0 \\
0 & 0 & 0 & 0 & b_{75} & b_{76} & b_{77} & b_{78} & b_{79} & 0 & 0 \\
0 & 0 & 0 & 0 & 0 & 0 & b_{87} & b_{88} & b_{89} & 0 & 0 \\
0 & 0 & 0 & 0 & 0 & 0 & b_{97} & b_{98} & b_{99} & b_{910} & 0 \\
0 & 0 & 0 & 0 & 0 & 0 & 0 & 0 & b_{109} & b_{1010} & 0 \\
0 & 0 & 0 & 0 & 0 & 0 & 0 & 0 & b_{119} & b_{1110} & -1
\end{pmatrix}
$$

$$A_5 = \begin{pmatrix} -1 & c_{12} & c_{13} & 0 & 0 & 0 & 0 & 0 & 0 & 0 & 0 \\ 0 & c_{22} & c_{23} & 0 & 0 & 0 & 0 & 0 & 0 & 0 & 0 \\ 0 & c_{32} & c_{33} & c_{34} & c_{35} & 0 & 0 & 0 & 0 & 0 & 0 \\ 0 & 0 & c_{43} & c_{44} & c_{45} & 0 & 0 & 0 & 0 & 0 & 0 \\ 0 & 0 & c_{53} & c_{54} & c_{55} & c_{56} & c_{57} & 0 & 0 & 0 & 0 \\ 0 & 0 & 0 & 0 & c_{65} & c_{66} & c_{67} & 0 & 0 & 0 & 0 \\ 0 & 0 & 0 & 0 & c_{75} & c_{76} & c_{77} & c_{78} & c_{79} & 0 & 0 \\ 0 & 0 & 0 & 0 & 0 & 0 & c_{87} & c_{88} & c_{89} & 0 & 0 \\ 0 & 0 & 0 & 0 & 0 & 0 & c_{97} & c_{98} & c_{99} & c_{910} & 0 \\ 0 & 0 & 0 & 0 & 0 & 0 & 0 & 0 & c_{109} & c_{1010} & 0 \\ 0 & 0 & 0 & 0 & 0 & 0 & 0 & 0 & c_{119} & c_{1110} & -1 \end{pmatrix}$$

$$X_3 = \begin{bmatrix} u_1 \\ u_2 \\ u_3 \\ u_4 \\ u_5 \\ u_6 \\ u_7 \\ u_8 \\ u_9 \\ u_{10} \\ u_{11} \end{bmatrix} \qquad X_4 = \begin{bmatrix} \theta_1 \\ \theta_2 \\ \theta_3 \\ \theta_4 \\ \theta_5 \\ \theta_6 \\ \theta_7 \\ \theta_8 \\ \theta_9 \\ \theta_{10} \\ \theta_{11} \end{bmatrix} \qquad X_5 = \begin{bmatrix} N_1 \\ N_2 \\ N_3 \\ N_4 \\ N_5 \\ N_6 \\ N_7 \\ N_8 \\ N_9 \\ N_{10} \\ N_{11} \end{bmatrix}$$

$$B3 =$$

$$-\frac{1}{525}\, r\left(48\,s\,N^5 + 8\,N\left(s\,N + 35\,u\right) + 5\,N\left(s\,N + 7\left(-4u + 3u\right)\right)\right)$$

$$-\frac{1}{525}\, r\left(5\,s\,N^7 + 48\,s\,N^5 + 8\,N\left(s\,N + 35\left(-u + u\right)\right) + N\left(8\,s\,N - 4\,s\,N - 35\left(3u - 4u + u\right)\right) + 5\,N\left(s\,N + 7\left(u - 4u + 3u\right)\right)\right)$$

$$-\frac{1}{525}\, r\left(5\,s\,N^7 + 48\,s\,N^5 + 8\,N\left(s\,N + 35\left(-u + u\right)\right) + N\left(8\,s\,N - 4\,s\,N - 35\left(3u - 4u + u\right)\right) + 5\,N\left(s\,N + 7\left(u - 4u + 3u\right)\right)\right)$$

$$-\frac{1}{525}\, r\left(5\,s\,N^7 + 48\,s\,N^5 + 8\,N\left(s\,N + 35\left(-u + u\right)\right) + N\left(8\,s\,N - 4\,s\,N - 35\left(3u - 4u + u\right)\right) + 5\,N\left(s\,N + 7\left(u - 4u + 3u\right)\right)\right)$$

$$-\frac{1}{525}\, r\left(5\,s\,N^7 + 48\,s\,N^5 - 280\,N\,u + N\left(8\,s\,N - 35\left(3u - 4u\right)\right)\right)$$

50

$$B4 = \left\{ \begin{array}{c}
-\dfrac{r\left(4\,N_2 - N_3\right)}{6\,(1+r)} - \dfrac{s\left(4\,Q+2\,t_2 - t_3\right)}{150\,(1+r)} \\[3mm]
-\dfrac{2\,G\left(0.\!\!\;\!+0.\!\!\;\!\left(3000+150\,s-s^3\right)\right)}{3\,(1+r)\,s^3} - \dfrac{2\,r\,N_3}{3\,(1+r)} - \dfrac{s\left(Q+8\,t_2 + t_3\right)}{75\,(1+r)} \\[3mm]
-\dfrac{r\left(-4\,N_2+3\,N_3\right)}{6\,(1+r)} - \dfrac{r\left(-3\,N_3+4\,N_4-N_5\right)}{6\,(1+r)} + \dfrac{s\left(Q-2\left(t_2-t_3\right)\right)}{150\,(1+r)} - \dfrac{s\left(4\,t_3+2\,t_4-t_5\right)}{150\,(1+r)} \\[3mm]
-\dfrac{2\,G\left(0.\!\!\;\!+0.\!\!\;\!\left(3000+150\,s-s^3\right)+0.\!\!\;\!\left(3000-150\,s+s^3\right)\right)}{3\,(1+r)\,s^3} + \dfrac{2\,r\left(N_3-N_5\right)}{3\,(1+r)} - \dfrac{s\left(t_3+8\,t_4+t_5\right)}{75\,(1+r)} \\[3mm]
-\dfrac{r\left(N_3-4\,N_4+3\,N_5\right)}{6\,(1+r)} - \dfrac{r\left(-3\,N_5+4\,N_6-N_7\right)}{6\,(1+r)} + \dfrac{s\left(t_3-2\left(t_4+2\,t_5\right)\right)}{150\,(1+r)} - \dfrac{s\left(4\,t_5+2\,t_6-t_7\right)}{150\,(1+r)} \\[3mm]
-\dfrac{2\,G\left(0.\!\!\;\!+0.\!\!\;\!\left(3000+150\,s-s^3\right)+0.\!\!\;\!\left(3000-150\,s+s^3\right)\right)}{3\,(1+r)\,s^3} + \dfrac{2\,r\left(N_5-N_7\right)}{3\,(1+r)} - \dfrac{s\left(t_5+8\,t_6+t_7\right)}{75\,(1+r)} \\[3mm]
-\dfrac{r\left(N_5-4\,N_6+3\,N_7\right)}{6\,(1+r)} - \dfrac{r\left(-3\,N_7+4\,N_8-N_9\right)}{6\,(1+r)} + \dfrac{s\left(t_5-2\left(t_6+2\,t_7\right)\right)}{150\,(1+r)} - \dfrac{s\left(4\,t_7+2\,t_8-t_9\right)}{150\,(1+r)} \\[3mm]
+\dfrac{2\,r\left(N_7-N_9\right)}{3\,(1+r)} - \dfrac{s\left(t_7+8\,t_8+t_9\right)}{75\,(1+r)} \\[3mm]
-\dfrac{r\left(N_7-4\,N_8+3\,N_9\right)}{6\,(1+r)} - \dfrac{r\left(-3\,N_9+4\,N_{10}\right)}{6\,(1+r)} + \dfrac{s\left(t_7-2\left(t_8+2\,t_9\right)\right)}{150\,(1+r)} - \dfrac{s\left(-Q+4\,t_9+2\,t_{10}\right)}{150\,(1+r)} \\[3mm]
\dfrac{2\,r\,N_9}{3\,(1+r)} - \dfrac{s\left(Q+t_9+8\,t_{10}\right)}{75\,(1+r)} \\[3mm]
-\dfrac{r\left(N_9-4\,N_{10}\right)}{6\,(1+r)} + \dfrac{s\left(t_9-2\left(2\,Q+t_{10}\right)\right)}{150\,(1+r)}
\end{array} \right\}$$

$$B5 = \left\{ \begin{array}{c}
\dfrac{b\,G\,(-30+s)\,s}{6\,a\,s^3} + \dfrac{r\left(100\,u_2-25\,u_3\right)}{150\,a} \\[3mm]
\dfrac{2\,r\left(25\,u_3\right)}{75\,a} \\[3mm]
\dfrac{r\left(-100\,u_2+75\,u_3\right)}{150\,a} + \dfrac{r\left(-75\,u_3+100\,u_4-25\,u_5\right)}{150\,a} \\[3mm]
\dfrac{2\,r\left(-25\,u_3+25\,u_5\right)}{75\,a} \\[3mm]
\dfrac{r\left(25\,u_3-100\,u_4+75\,u_5\right)}{150\,a} + \dfrac{r\left(-75\,u_5+100\,u_6-25\,u_7\right)}{150\,a} \\[3mm]
\dfrac{2\,r\left(-25\,u_5+25\,u_7\right)}{75\,a} \\[3mm]
\dfrac{r\left(25\,u_5-100\,u_6+75\,u_7\right)}{150\,a} + \dfrac{r\left(-75\,u_7+100\,u_8-25\,u_9\right)}{150\,a} \\[3mm]
\dfrac{2\,r\left(-25\,u_7+25\,u_9\right)}{75\,a} \\[3mm]
\dfrac{r\left(25\,u_7-100\,u_8+75\,u_9\right)}{150\,a} + \dfrac{r\left(-75\,u_9+100\,u_{10}\right)}{150\,a} \\[3mm]
\dfrac{2\,r\left(-25\,u_9\right)}{75\,a} \\[3mm]
\dfrac{r\left(25\,u_9-100\,u_{10}\right)}{150\,a}
\end{array} \right\}$$

The equilibrium conditions are

$$R_3^1 + R_1^2 = 0, \qquad\qquad R_3^2 + R_1^3 = 0,$$

$$R_3^3 + R_1^4 = 0, \qquad\qquad R_3^4 + R_1^5 = 0,$$

$$Q_3^1 + Q_1^2 = 0, \qquad\qquad Q_3^2 + Q_1^3 = 0,$$

$$Q_3^3 + Q_1^4 = 0, \qquad\qquad Q_3^4 + Q_1^5 = 0,$$

$$S_3^1 + S_1^2 = 0, \qquad\qquad S_3^3 + S_1^4 = 0,$$

$$S_3^4 + S_1^5 = 0, \qquad\qquad\qquad\qquad\qquad 4.4$$

Solving these coupled global matrices for temperature, concentration and velocity (4.1) – (4.4) respectively and using the iteration procedure we determine the unknown global nodes through which the temperature, concentration and velocity at different radial intervals at any arbitrary axial cross sections are obtained.

2.5. DISCUSSION:

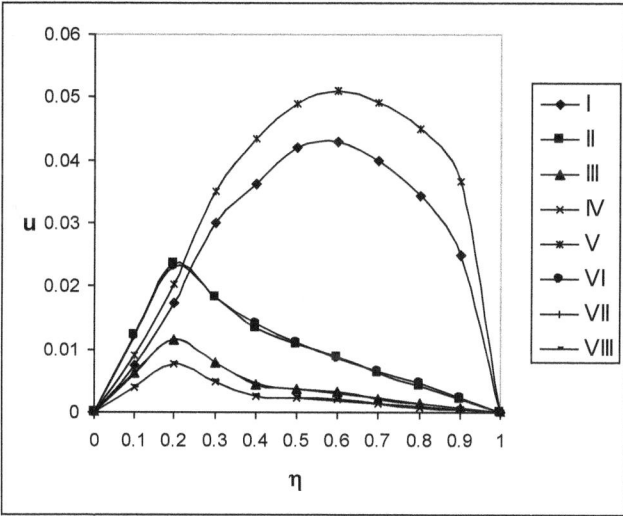

Fig. 1 : Variation of u with D^{-1}

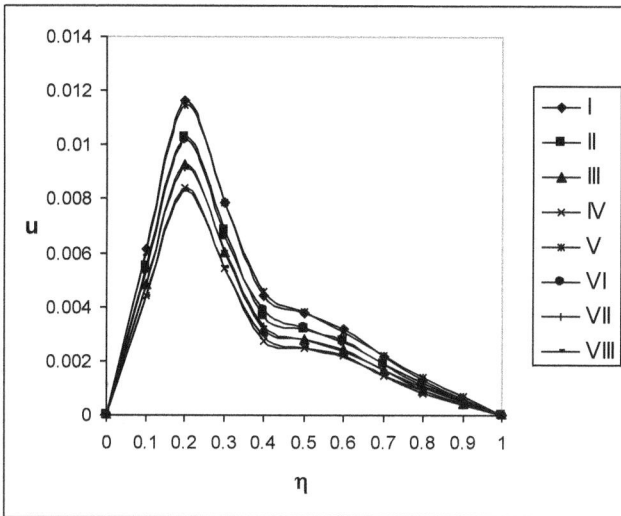

Fig. 2 : Variation of u with H

53

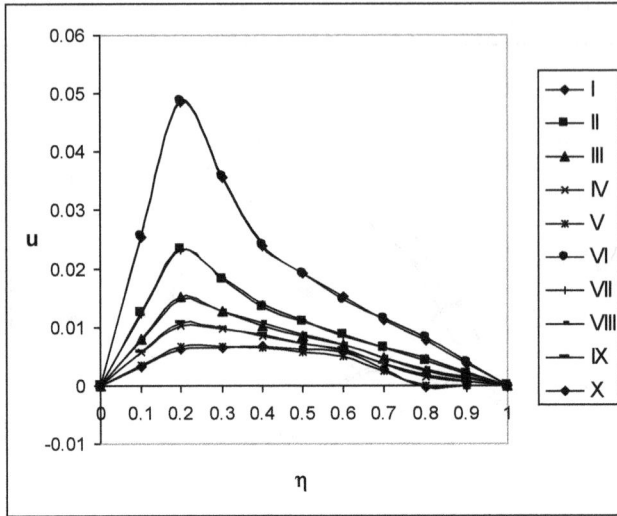

Fig. 3 : Variation of u with R

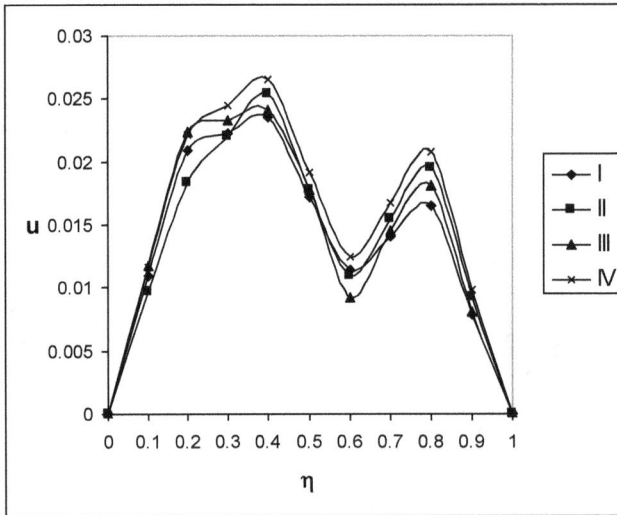

Fig. 4 : Variation of u with γ

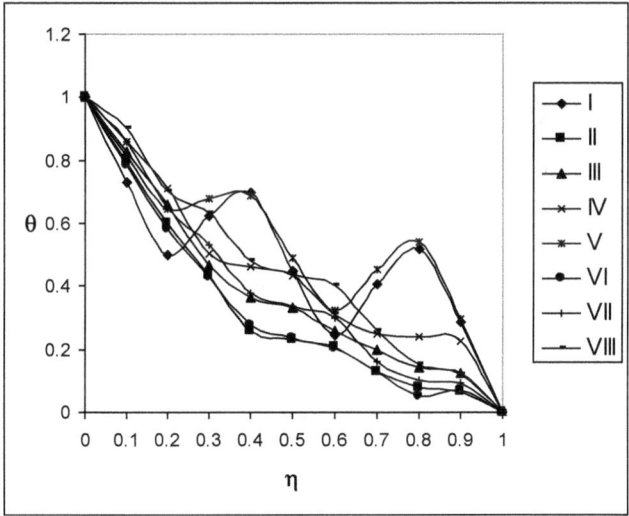

Fig. 5 : Variation of θ with D⁻¹

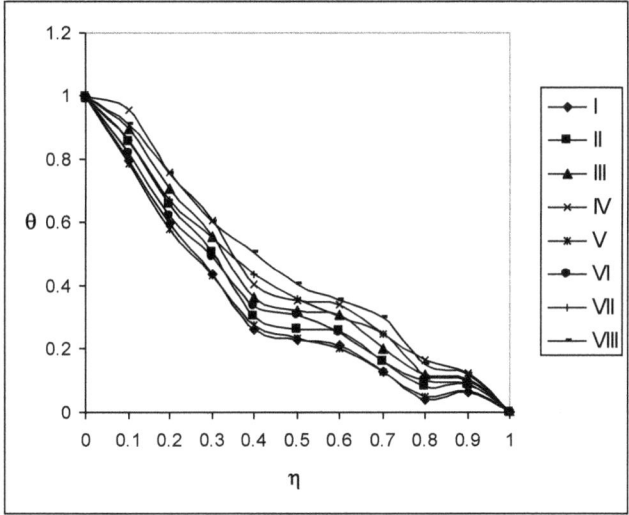

Fig. 6 : Variation of θ with H

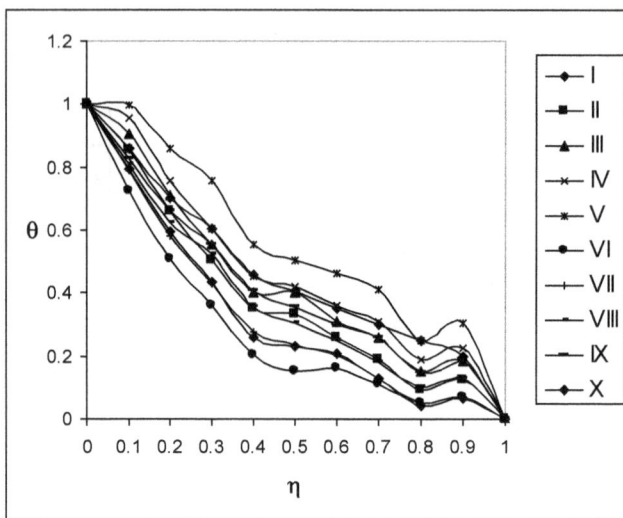

Fig. 7 : Variation of θ with R

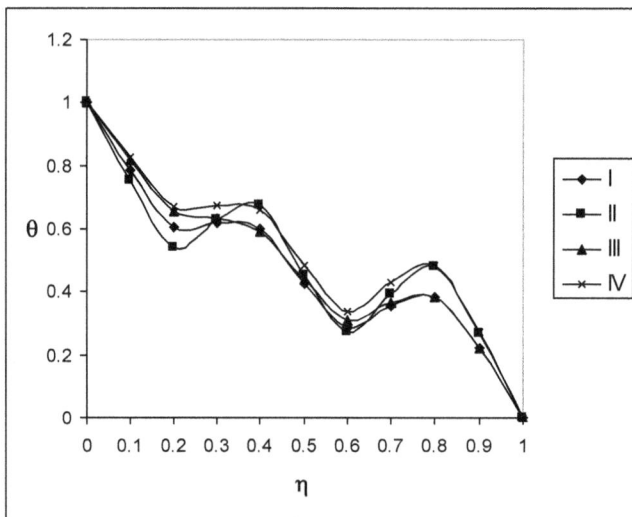

Fig. 8 : Variation of θ with γ

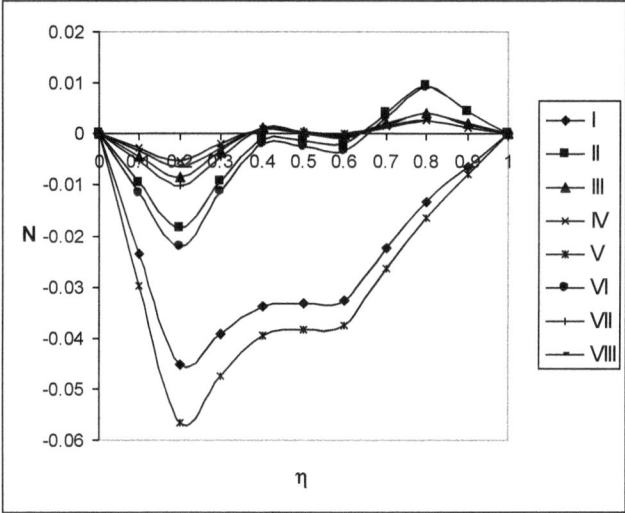

Fig. 9 : Variation of N with D^{-1}

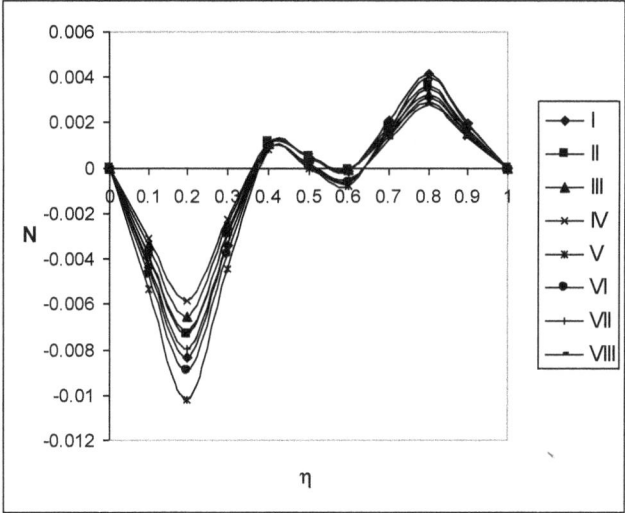

Fig. 10 : Variation of N with H

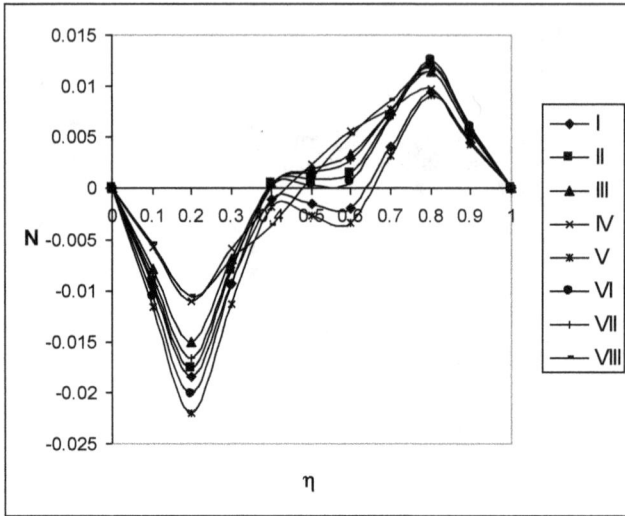

Fig. 11 : Variation of N with R

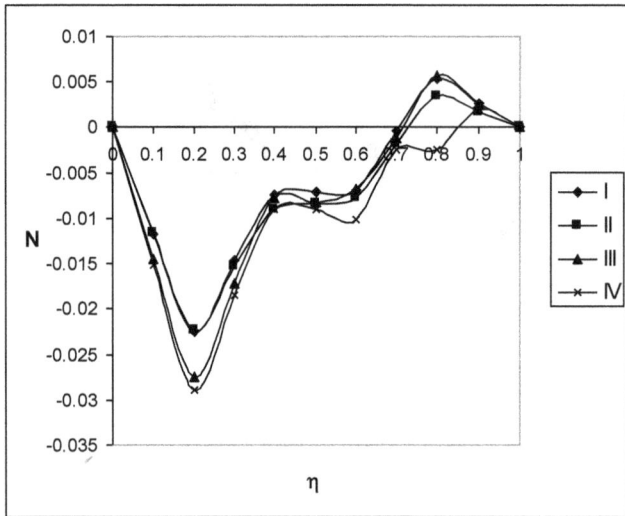

Fig. 12 : Variation of N with γ

The system of equations with boundary conditions velocity, micro rotation, temperature distribution is obtained by employing the Galarkin finite element technique. The Prandtl number (Pr) and micro polar material constant a,b,c are taken to be constant at 0.733, 1.0, 0.001 and 0.1 respectively whereas the effect of important parameters namely, the micro polar parameter R, the cross flow Reynolds number Re, Darcy parameter D^{-1}, local Hartmann number M, has been studied for these functions and corresponding profiles are shown in 1-12.

Fig.1 exhibits the variation of velocity function u with Darcy parameter D^{-1} for a prescribed value of micro polar parameter R_e = 1 and Hartmann number M=5 profiles are shown both for positive Re(=2) i.e., for injection at $\eta = 0$ and suction E = 1 and negative Re (= - 2) i.e., for suction at $\eta = 0$ and injection at $\eta=1$. The velocity profiles are parabolic in nature and attaint he maximum near the middle for $D^{-1} = 0$ and for D^{-1} different from 0 the maximum occurs at $\eta = 0.2$. This is trace for both positive and negative values of Re. It is 0bserved that the velocities are higher for injection has compared to suction at $\eta = 0$ it is found that lesser the permeability of the porous medium larger the velocity in the vicinity of $\eta=0$ and smaller the velocity in the remaining flow region and for further lowering of the permeability smaller the velocity in the entire flow region for both injection and suction. The velocity continuously decreases with an increase in the Hartmann number. The same pattern is observed for both values of cross flow Reynolds number Re. This indicates that the fluid velocity can be reduced by an increase in the magnetic field.

A similar phenomenon is observed in (Gorla Takhar Staouti) the variation in velocity with reference to variation in micro polar parameter R for constant values of R and R_e is presented in fig.3. The results are given both for positive and negative cross flow Reynolds number Re. It is observed that the velocity values for suction at $\eta = 0$ as the micro polar parameter R increases the velocity continuously decreases the maximum being shifted to $\eta = 0$. This situation is valid for both positive Re and negative Re. This is however different as compared to viscous flow which

59

corresponds to $R_e = 0$. It is worth noting to observe this behavior of micro polar flow. The boundary layer thickness as expected continuously decreases with increasing micro concentration. The variation of velocity with heat source parameter γ is presented in fig.4. It is observed that an increase in γ continuously depreciates the velocity with the maximum accruing at $\eta = 1.4$.

The non-dimensional temperature distribution is shown in figs. 5-8. For different values at D^{-1}, M, Re, &γ. The profiles for both positive and negative values of Re. The profile of the temperature gradually reduces from a prescribed values 1 on $\eta = 0$ to attaint he values $\eta = 0$ and $\eta = 1$. It is noticed that the temperature is positive for all values of D^{-1}, M, R, Re and γ lesser the permeability of the porous medium larger the temperature in the vicinity of $\eta = 0$ and smaller in the remaining region and for further lowering of permeability larger the temperature in entire flow region in the injection case. In the case of suction depreciates with $D^{-1} \leq 10^2$ and enhances with higher $D^{-1} \geq 3 \times 10^2$. The variation of θ with magnetic Hartmann number is shown if fig. 6. It is observed that the temperature in the case of injection is slightly higher than that of suction and increase in the Hartmann number M continuously enhances the temperature everywhere in the flow region. The variation of θ with micro polar parameter R is shown in fig.7 an increase the micro polar parameter R continuously enhances the temperature in the entire flow region for both negative and positive values of Re. In fig.8 represents the variation of θ with R. It is found that in the injection case an increase in γ depreciates the temperature in the flow region and enhances it in the suction case.

The micro rotation is exhibited in figs. 9-12 for a different variations D^{-1}, H, R and γ. The effect of Darcy parameter D^{-1} on micro rotation in the micro polar parameter R_e kept fixed at three is shown in fig.9. It is found that the values of micro rotation are negative in the first half where as in the second half are positive. Thus show in the reversed rotation near the two boundaries moreover for section these are higher numerically as compared to injection at $\eta=0$ an increase in the magnetic field leads to a decrease in the micro rotation except in the central case where we find an enhancement in the micro rotation for $M \geq 15$ while for higher $M \geq 20$ the micro rotation continuously decrease in the entire flow region. The intensity of the magnetic field thus can be used for decreasing in the angular rotation especially in the suspension flow commonly raised in lubrication problems in fig.10. Fig.11 represents the variation of micro rotation with micro polar parameter R for both suction and injection it is observed that the micro rotation depreciates in the first half and enhances in second half with increasing R for Re=2 while for Re= -2 the micro rotation depreciates except in the vicinity of $\eta=1$ for $R \leq s$ for higher R=3 the micro rotation enhances except in the vicinity of $\eta=0$ and for still higher R=5 the micro rotation experiences is depreciation in the region adjust to the walls $\eta=0$ and 1 and enhances in the central flow region. Fig.12 represents the variation of micro rotation with increase in γ it is found that in the injection case an increase in γ accurate the micro rotation except in an arrow region obtaining $\eta=0$ while in the suction case the micro rotation enhances in the first half depreciates in the second half in the magnitude.

A skin friction at the boundaries $y = \pm 1$ has been given for Re $= 2$ and -2 in the tables 1 & 2 which show the variation with respect to Darcy parameter D^{-1}, Hartmann number M, micro polar parameter R and γ, from the numerical results obtained as given tables 1 & 2 it is found that the skin friction for injection decreases within increase the Darcy parameter D^{-1} Hartmann number M, keeping the micro parameter R fixed, thus causing resistance. Also it decreases with micro polar parameter R at $y = \pm 1$ and at $y = -1$ it decreases with $\gamma \leq 5$ and enhances with higher $R \geq 10$ but for the increasing R reduced the skin friction at both the place. A similar behavior is noticed with respect to this parameter in the suction cases. A Nusselt number (Nu) with represents the rate at $y = \pm 1$ of heat transfer $y=\pm 1$ is shown in tables 3 & 4 for a different parametric values. It is found in both suction & injection cases. The rate of heat transfer decreases with D^{-1} & M. Thus lesser the permeability of the porous medium (or) higher the Lorentz force smaller the rate of heat transfer at both the walls also the rate of heat transfer experiences and enhancement with increasing micro polar parameter R. This implies that the magnetic field as well as porous medium and micro concentration are the important parameters in reducing/enhancing heating effects and can be used for controlling the rate of heat transfer which as desired in many MHD application. The couple stress at $y = \pm 1$ are calculated for different values G, D^{-1}, M, R and γ. It is observed that lesser the permeability of the porous medium (or) Lorentz force (or) micro polar parameter R(or) higher the micro concentration smaller the couple stress at $y = 1$ & larger couple stress at $y = -1$ an increasing R≤3 enhances the couple stress at $y = 1$ and reduces it is $y = -1$ and reversed effect is observed in the behavior of couple stress with higher $\gamma \geq 5$.

REFERENCE:

[1] ARIMAN, T. TURK, M.A. SYLVESTER, N.D.: Applications of

Micro continuum fluid mechanics. Int. J. Eng, sci. 12, 273-293 (1974).

[2] AGARWAL, R.S. C. DHANAPAL, Numerical solution of free convection

Micro polar fluid flow between two parallel porous vertical plates, Int. J. Eng.

sci. 26(1988) 1247-1255.

[3] BALARAM, M. SASTRY, N.V.K; micro polar free convection. Int. J. Heat

Mass Transfer 16, 437-441 (1973).

[4] ERIGEN, A.C. micro rotation field theories II Fluent, media, springer – veriag,

New York, 2001.

[5] GORLA, R.S.N, TAKHAR, H.S. A. SLAOUTI, magneto hydrodynamic free

Convection boundary layer flow of a thermo micro polar fluid over a vertical

plate, Int. J. Eng. sci. 36 (1998) 315-327.

[6] JENA, S.K. MATHUS, M.N. similarly solutions for laminar free convection

flow of a thermo micro polar fluid past a non-isothermal vertical flat plate. Int.

J. Eng. sic, 19, 1431-1439 (1981).

[7] KASIVISHWANATHAN S.R, M.V. GANDHI, A class of exact solutions for

the magneto hydrodynamic flow of a micro polar fluid. Int. J. Eng. sci, 30

(1992) 409-417.

[8] MOHAMMADEIN, A.A, GORLA, R.S.R. Effects of transverse magnetic field

on a mined convection in a micro polar fluid on a horizontal plate with vectored mass transfer, Acta. Mech. 118 (1966) 1-12.

[9] RAMA BHARAGA[a]. L – KUMA[a], H.S. TAKHAR. Numerical solution of

free convection MHD micro polar fluid flow between two parallel porous plates

Int. J. Eng. Sci 41(2003) 123-136.

CHAPTER - III

NUMERICAL STUDY OF CONVECTIVE HEAT TRANSFER FLOW OF A MICROPOLAR FLUID THROUGH A POROUS MEDIUM IN A CYLINDRICAL ANNULUS

3.1. INTRODUCTION:

In recent years there has been considerable advancement in the study of free convection in a porous annulus because of its natural occurrence and of its importance in many branches of science and engineering. This is of fundamental importance to a number of technological applications, such as underground disposal of radioactive waste materials, cooling of nuclear fuel in shipping flasks and water filled storage bays, regenerative heat exchangers containing porous materials and petroleum reservoirs, burying of drums containing heat generating chemicals in the earth, storage of agricultural products, ground water flow modeling, nuclear reactor assembly, thermal energy storage tanks, insulation of gas cooled reactor vessels, high performance insulation for building, and cold storage, free convective heat transfer inside a vertical annulus filled with isotropic porous medium has been studied by many researchers, notable among them are Havstad et. al., [7], Reda [15], Prasad et. al., [12,13], Prasad et. al., [14], Hickon et. al., [8], and Shivakumara et. al., [16], they concluded that radius ratio and Rayleigh number influence the heat and fluid flow significantly.

Eringen [5] presented the theory of micro polar fluid which explains adequately the internal characteristics of the substructure particles subject to rotations and deformations, Ariman [1] Sylvester et. al., [17] and Ariman et. al., [2] confirmed that the micro polar fluid serves a better model for animal blood. The liquid crystals, suspension solutions and certain polymeric fluids consisting of randomly oriented barnlike elements of dumbbell molecules, behave as micro polar fluid. Kazakai and Ariman [10] introduced the heat conducting micro polar fluid and investigated the flow between two parallel plates. Eringen [6] extended the theory further to include heat conduction and viscous dissipation. The application of this theory may be searched in biomechanics.

Berman [4] has studied the flow of a viscous liquid under a constant pressure gradient between two concentric cylinders at rest. Inger [9] solved the problem when the walls are permeable and the outer cylinder is sliding with a constant axial velocity relative to the stationery inner one. Mishra and Acharya [11] examined the elastico viscous effects in the flow considered by Inger [9]. Sastri and Maiti [18] solved the problem of combined convective flow of micro polar fluid in an annulus.

The problem of massive blowing into aerodynamic body flow fields is a complex 1 for a realistic flow configuration. However the study the simplified flow models exhibits some of the essential physical features involving the interaction of blowing with a shear flow. The classical coutte – poiseylle shear flow is an idealized model for this purpose to investigate to micro polar structures at least theoretically on such a flow Agarwal and Dhanpal [3] have analyzed the connective micro polar fluid flow and heat transfer between two concentric porous circular cylinders when the outer cylinder moves parallel to itself with a constant velocity. A problem solved by Inger [9] in viscous fluids.

In this chapter we make an investigation of connective heat transfer flow micro polar fluid in a cylindrical annulus between the porous concentric cylinders $r = 0$ and $r = b$. By complying Galarikine finite element analysis with triangular elements the transport equations of linear momentum, angular momentum and energy are solved to obtain a velocity micro concentration and temperature distributions. The stress, rate of heat transfer on the couple stress on the cylinders are numerically evaluated for different values of G, λ, D^{-1}, Δ.

3.2. FORMULATION OF THE PROBLEM (m = 0, Ec = 0):

Consider the steady motion of an incompressible micro polar fluid through an annulus of two infinitely long porous circular cylinders of radii a and b (a – b = h > 0) respectively. The fluid is injected through the inner cylinder with arbitrary radial velocity u_b and in view of continuity, also flows outward through the moving cylinder with a radial velocity u_a. The cylindrical polar co-ordinate system (γ, θ, z) with z co-ordinate along the axis of the cylinders is chosen to specify the problem.

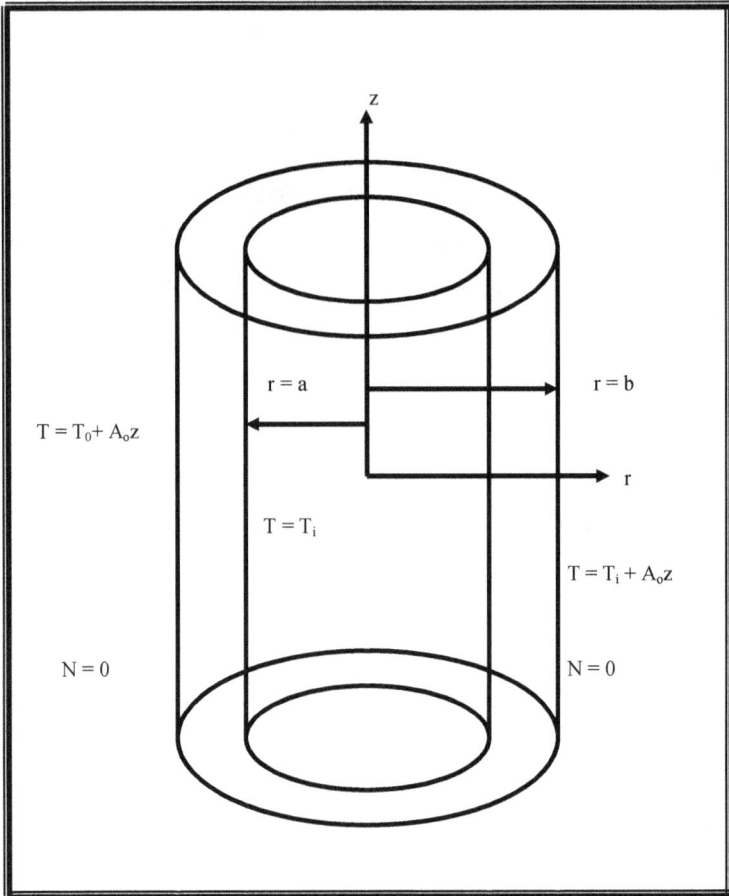

The velocity and micro rotation are taken in the form

$$v_r = u(r), \quad v_\theta = v = 0, \quad v_z = w(r)$$

$$v_r = 0, \quad v_\theta = v = (r), \quad v_z = 0 \qquad\qquad 2.1$$

The basic equations are:

Continuity:
$$\frac{\partial u}{\partial r} + \frac{u}{r} = 0 \qquad\qquad 2.2$$

Momentum:

$$\rho u \frac{\partial u}{\partial r} = -\frac{\partial p}{\partial r} + (\mu + k)\left(\frac{\partial^2 u}{\partial r^2} + \frac{1}{r}\frac{\partial u}{\partial r} - \frac{u}{r^2}\right) \qquad\qquad 2.3$$

$$\rho u \frac{\partial w}{\partial r} = -\frac{\partial p}{\partial z} + (\mu + k)\left(\frac{\partial^2 w}{\partial r^2} + \frac{1}{r}\frac{\partial w}{\partial r}\right) + \frac{k}{r}\frac{\partial}{\partial r}(rN) \qquad\qquad 2.4$$

First stress momentum:

$$\rho j u \frac{\partial N}{\partial r} = r\left(\frac{\partial^2 N}{\partial r^2} + \frac{1}{r}\frac{\partial N}{\partial r} - \frac{N}{r^2}\right) - k\frac{\partial w}{\partial r} - 2kN \qquad\qquad 2.5$$

Energy:

$$\rho C_p w \frac{\partial T}{\partial z} = k_f\left(\frac{\partial^2 T}{\partial r^2} + \frac{1}{r}\frac{\partial T}{\partial r}\right) \qquad\qquad 2.6$$

Where T is the temperature, N is the micro rotation ρ the density, p the pressure, j the micro inertia, C_p the specific heat, k_f the thermal conductivity, μ the fluid viscosity, k is the material constants.

at   r = b,   u = ub,   w = 0,   v = 0, $T = T_0 + A_0\,z$

at   r = a,   u = ua,   w = w₀,   v = 0, $T = T_i + A_0\,z$ 2.7

Where the fluid is assumed to adhere to the solid boundaries. A_0 is the constant of proportionality and T_0, T_1 are the constant wall temperatures of inner and outer cylinders temperatures of inner and outer cylinders respectively at z = 0.

The integration of (2.2) yields

        ur = C, a constant   2.8

$$\Rightarrow \ r\,u = a\,u_a = b\,u_b \ \Rightarrow \ u = \frac{a\,u_a}{r}$$

Also in view of the boundary condition on temperature, we may write

$$T = T_0 + A_\theta\,(z) + \theta(r) \qquad\qquad 2.9$$

Introducing the non-dimensional variables r', w', p', θ and N as

$$r' = \frac{r}{a}, \ w' = \frac{w}{a}, \ p' = \frac{p}{\rho\,u^2\big/a^2}, \ \theta = \frac{T-T_0}{T_i - T_0}, \ N' = \frac{(\mu+k)\,w}{\rho u^2 a^2} \qquad 2.10$$

The governing equations in the non-dimensional form are

$$\frac{d^2 w}{dr^2} + \left(1 - \frac{\lambda}{1+\Delta}\right)\frac{1}{r}\frac{dw}{dr} = -\pi_1 + D_1^{-1} - G\theta - \frac{\Delta_1}{r}\frac{d}{dr}(r\,N) \qquad 2.11$$

$$\frac{d^2 N}{dr^2} + (1 - \lambda A)\frac{1}{r}\frac{dN}{dr} - \left(\frac{1}{r^2} - \frac{2\Delta}{A}\right)N = \frac{\Delta}{A}\frac{1}{r}\frac{dw}{dr} \qquad 2.12$$

$$Pr\,N_T\,w = \frac{d^2\theta}{dr^2} + \frac{1}{r}\frac{d\theta}{dr} \qquad\qquad 2.13$$

Where $\quad \Delta = \dfrac{k}{\mu}\qquad\qquad$ (Micro polar parameter)

$$A = \frac{r}{\mu J}\qquad\qquad \text{(Micro polar parameter)}$$

$$A_1 = \frac{\beta}{\mu a^2}\qquad\qquad \text{(Micro polar parameter)}$$

$$\lambda = \frac{a\,u_a}{v}\qquad\qquad \text{(Suction parameter)}$$

$$D^{-1} = \frac{L^2}{k}\qquad\qquad \text{(Darcy parameter)}$$

$$G = \frac{\beta g a^3 \Delta T}{v^2}\qquad \text{(Grashof number)}$$

70

$$G_1 = \frac{G}{1+\Delta}$$

$$\Delta_1 = \frac{\Delta}{1+\Delta}$$

$$Pr = \frac{\mu C_p}{k_f} \qquad \text{(Prandtl number)}$$

The boundary conditions are

$$w = 0, \qquad \theta = 1, \qquad N = 0 \qquad \text{on} \qquad r = 1$$

$$w = 0, \qquad \theta = 0, \qquad N = 0 \qquad \text{on} \qquad r = s$$

$$E_w^k = \frac{d}{dr}\left(r\,\frac{dw}{dr}\right)^k - \left(\frac{\lambda}{(1+\Delta)}\right)\frac{dw^k}{dr} + \pi r D_1^{-1} w r^k + G_1 \theta^k r + \Delta_1 \frac{d}{dr}(r N^k) \quad 3.1$$

$$E_\theta^k = \frac{d}{dr}\left(\frac{r d\theta^k}{dr}\right) - r\, Pr\, N_T\, w^k \qquad\qquad 3.2$$

$$E_N^k = \frac{d}{dr}\left(r\,\frac{dN^k}{dr}\right) - (\lambda A)\frac{dN^k}{dr} - \left(\frac{1}{r} - \frac{2\Delta r}{A}\right)N^k - \frac{\Delta}{A}\frac{dw^k}{dr} \qquad 3.3$$

Where w^k, θ^k and N^k are values of w, θ and N in the arbitrary element e_k. These are expressed as linear combinations in terms of respective local modal values.

$$w^k = w_1^k \psi_1^k + w_2^k \psi_2^k + w_3^k \psi_3^k = \sum_{i=1}^{3} w_i^k \psi_i^k$$

$$\theta^k = \theta_1^k \psi_1^k + \theta_2^k \psi_2^k + \theta_3^k \psi_3^k = \sum_{i=1}^{3} \theta_i^k \psi_i^k$$

$$N^k = N_1^k \psi_1^k + N_2^k \psi_2^k + N_3^k \psi_3^k = \sum_{i=1}^{3} N_i^k \psi_i^k$$

where $\psi_1^k, +\psi_2^k, \ldots$ etc are Lagrange's quadratic polynomials.

$$\int_{r_A}^{r_B} r \frac{dw^k}{dr} \frac{d\psi^k}{dr} dr - \frac{\lambda}{1+\Delta} \int_{r_A}^{r_B} \frac{dw^k}{dr} \psi^k dr + A \int_{r_A}^{r_B} \psi^k dr + G_1 \int_{r_A}^{r_B} \theta^k \psi^k dr$$

$$+ \Delta_1 \int_{r_A}^{r_B} r N^k \psi^k dr = Q_{2,j}^k + Q_{1,j}^k \qquad 3.4$$

where

$$- \theta_{1,J}^k = \left(\left(\frac{dw^k}{dr} \right) r \psi_j^k \right)_{r A_1} + \Delta \left(N^k r \psi_j^k \right) r A$$

$$- \theta_{2,J}^k = \left(\left(\frac{dw^k}{dr} \right) r \psi_j^k \right)_{r_B} + \Delta \left(N^k r \psi_j^k \right) r B$$

$$\int_{r_A}^{r_B} \frac{d\theta^k}{dr} \frac{d\psi^k}{dr} dr = Nt \ Pr \int_{r_A}^{r_B} r \psi^k \ \psi_j^k \ dr + R_{2j}^k + R_{1j}^k \qquad 3.5$$

where

$$- R_{1,J}^k = \left(\left(\frac{d\theta^k}{dr} \right) (r \psi_j^k) \right)_{rA} ,$$

$$- R_{2,J}^k = \left(\left(\frac{d\theta^k}{dr} \right) (r \psi_j^k) \right)_{rB}$$

$$\int_{r_A}^{r_B} r \frac{dN^k}{dr} \frac{d\psi_j^k}{dr} dr = \lambda A \int_{r_A}^{r_B} N^k \frac{d\psi_j^k}{dr} + \frac{2\Delta r}{A} \int_{r_A}^{r_B} \frac{1}{r} N_k \ \psi_j^k \ dr$$

$$+ \frac{\Delta}{A} \int_{r_A}^{r_B} w^k \frac{d\psi_j^k}{dr} = S_{1j}^k + S_{2j}^k \qquad 3.6$$

$$- S_{1J}^k = \left(\left(\frac{dN^k}{dr} \right) (r \psi_j^k) \right)_{rA} ,$$

$$- R_{2J}^k = \left(\left(\frac{dN^k}{dr} \right) (r \psi_j^k) \right)_{rB}$$

Expressing w^k, θ^k, N^k in terms of local nodal values in (3.4) – (3.6) we obtain

$$\sum_{i=1}^{3} w^k \int_{r_A}^{r_B} r \frac{d\psi_i^k}{dr} \frac{d\psi_j^k}{dr} \, dr - \frac{\lambda}{1+\Delta} \sum_{i=1}^{3} w^k \int_{r_A}^{r_B} \frac{d\psi_i^k}{dr} \psi_j^k \, dr + \sum_{i=1}^{3} w^k \int_{r_A}^{r_B} \psi_i^k \psi_j^k \, dr$$

$$+ \sum_{i=1}^{3} G_i \theta^k \int_{r_A}^{r_B} \psi_i^k \psi_j^k \, dr + \Delta_i \sum_{i=1}^{3} N^k \int_{r_A}^{r_B} r \psi_i^k \psi_j^k \, dr = Q_{1,J}^k + Q_{2,J}^k \qquad 3.7$$

$$\sum_{i=1}^{3} \theta_i^k \int_{r_A}^{r_B} r \frac{d\psi_i^k}{dr} \frac{d\psi_j^k}{dr} \, dr - N_t \, Pr \sum_{k=1}^{3} w_i^k \int_{r_A}^{r_B} r \psi_i^k \psi_j^k \, dr = R_{1,j}^k + R_{2,j}^k \qquad 3.8$$

$$\sum_{i=1}^{3} N_i^k \int_{r_A}^{r_B} r \frac{dw_i^k}{dr} \frac{dw_j^k}{dr} \, dr - \lambda A \sum_{i=1}^{3} N_i^k \int_{r_A}^{r_B} \frac{d\psi_i^k}{dw} \psi_j^k \, dr$$

$$- \sum_{i=1}^{3} N_i^k \int_{r_A}^{r_B} \frac{\psi_i^k}{r} w_j^k \, dr - \frac{\Delta}{A} \sum_{i=1}^{3} w_i^k \int_{r_A}^{r_B} \frac{d\psi_i^k}{dw} \psi_j^k = S_{2,j}^k + S_{1,j}^k \qquad 3.9$$

Choosing different ψ_j^k's corresponding to each element e_k in the equation (3.7) yields a local stiffness matrix of order 3×3 in the form

$$\left(f_{iJ}^k\right)\left(w_i^k\right) - \delta G\left(h_{iJ}^k\right)\left(\theta_i^k + NN_i^k\right) + \delta D^{-1}\left(m_{iJ}^k\right)\left(w_i^k\right) + \delta^2 A\left(n_{iJ}^k\right)\left(w_i^k\right) \qquad 3.10$$

Likewise the equation (3.8) & (3.9) gives rise to stiffness matrices

$$\left(e_{iJ}^k\right)\left(\theta_i^k\right) - N_t \, Pr \left(w_i^k\right) = R_{2,J}^k + R_{1,J}^k \qquad 3.11$$

$$\left(l_{iJ}^k\right)\left(N_i^k\right) - \lambda A\left(t_{iJ}^k\right)\left(N_i^k\right) - \frac{\Delta}{A}\left(w_i^k\right) = S_{2,J}^k + S_{1,J}^k \qquad 3.12$$

where

$$\left(f_{iJ}^k\right), \left(g_{iJ}^k\right), \left(m_{iJ}^k\right), \left(n_{iJ}^k\right), \left(e_{iJ}^k\right), \left(l_{iJ}^k\right) \text{ and } \left(t_{iJ}^k\right) \text{ are} \qquad 3\times3 \qquad \text{matrices} \qquad \text{and}$$

$$V_j^k = -P_1 \int_{r_A}^{r_B} r \psi_i^k \psi_j^k \, dr \quad \text{and} \quad \left(Q_{2J}^k\right), \left(Q_{1J}^k\right), \left(R_{2J}^k\right), \left(R_{1J}^k\right), \left(S_{2J}^k\right) \text{ and } \left(S_{1J}^k\right) \quad \text{are} \quad 3\times1 \quad \text{column}$$

matrices and such stiffness (3.10) – (3.12) in terms of local nodes in each element are assembled using inter element continuity and equilibrium conditions to obtain the coupled global matrices in terms of the global nodal values of u, θ & C. In case we choose n-quadratic elements then the global matrices are of order 2n+1. The ultimate coupled global matrices are solved to determine the unknown global nodal values of the velocity, temperature and concentration in fluid region. In solving these global matrices an iteration procedure has been adopted in include the boundary and effects in the porous region.

In fact, the non-linear term arises in the modified Brinkman Linear momentum equation (3.4) of the porous medium. The iteration procedure in taking the global matrices as follows, we split the square term into a product term and keeping one of them say w_i's under integration. The other is expanded in terms of local nodal values as in (3.6), resulting in the corresponding coefficient matrix (n_{ij}^k, S) in (3.8), whose coefficients involve the unknown w_i's To evaluated (3.9) to begin with choose the initial global nodal values of u_i's as zeros in the zeroth approximation, we evaluate w_i's, θ_i's and N_i's in the usual procedure mentioned earlier. Later choosing these values of u_i's, N_i's. In the second iteration, we substitute for w_i's the first order approximation of and w_i's and the first approximation of θ_i's and N_i's obtain second order approximation. This procedure is repeated till the consecutive values of w_i's, θ_i's and N_i's differ by a reassigned percentage, for computation purpose we choose five elements in flow region.

The shape function corresponding to

$$\psi_1^1 = \frac{50\left(-1+r-\dfrac{S}{5}\right)\left(-1+r-\dfrac{S}{10}\right)}{S^2}$$

$$\psi_2^1 = \frac{100\left(-1+r\right)\left(-1+r-\dfrac{S}{5}\right)}{S^2}$$

$$\psi_3^1 = \frac{50\left(-1+r\right)\left(-1+r-\dfrac{S}{10}\right)}{S^2}$$

$$\psi_1^2 = \frac{50\left(-1+r-\dfrac{2S}{5}\right)\left(-1+r-\dfrac{3S}{10}\right)}{S^2}$$

$$\psi_2^2 = \frac{-100\left(-1+r-\dfrac{2S}{5}\right)\left(-1+r-\dfrac{S}{5}\right)}{S^2}$$

$$\psi_3^2 = \frac{50\left(-1+r-\dfrac{3S}{5}\right)\left(-1+r-\dfrac{s}{5}\right)}{S^2}$$

74

$$\psi_1^3 = \frac{50\left(-1+r-\dfrac{3S}{5}\right)\left(-1+r-\dfrac{S}{2}\right)}{S^2}$$

$$\psi_2^3 = \frac{100\left(-1+r-\dfrac{3S}{5}\right)\left(-1+r-\dfrac{2S}{5}\right)}{S^2}$$

$$\psi_3^3 = \frac{50\left(-1+r-\dfrac{S}{2}\right)\left(-1+r-\dfrac{2S}{5}\right)}{S^2}$$

$$\psi_1^4 = \frac{50\left(-1+r-\dfrac{4S}{5}\right)\left(-1+r-\dfrac{7S}{10}\right)}{S^2}$$

$$\psi_2^4 = \frac{-100\left(-1+r-\dfrac{4S}{5}\right)\left(-1+r-\dfrac{3S}{5}\right)}{S^2}$$

$$\psi_3^4 = \frac{50\left(-1+r-\dfrac{7S}{10}\right)\left(-1+r-\dfrac{3S}{5}\right)}{S^2}$$

$$\psi_1^5 = \frac{50(-1+r-S)\left(-1+r-\dfrac{9S}{10}\right)}{S^2}$$

$$\psi_2^5 = \frac{-100(-1+r-S)\left(-1+r-\dfrac{4S}{5}\right)}{S^2}$$

$$\psi_3^5 = \frac{50\left(-1+r-\dfrac{9S}{10}\right)\left(-1+r-\dfrac{4S}{5}\right)}{S^2}$$

3.4. STIFFNESS MATRICES:

The global matrix for θ is

$$A_3 X_3 = B_3 \qquad\qquad 4.1$$

The global matrix for N is

$$A_4 X_4 = B_4 \qquad\qquad 4.2$$

The global matrix w is

$$A_5 X_5 = B_5 \qquad\qquad 4.3$$

where

$$A3 = \begin{pmatrix}
a_{11} & a_{12} & a_{13} & 0 & 0 & 0 & 0 & 0 & 0 & 0 & 0 \\
0 & a_{22} & a_{23} & 0 & 0 & 0 & 0 & 0 & 0 & 0 & 0 \\
0 & a_{32} & a_{33} & a_{34} & a_{35} & 0 & 0 & 0 & 0 & 0 & 0 \\
0 & 0 & a_{43} & a_{44} & a_{45} & 0 & 0 & 0 & 0 & 0 & 0 \\
0 & 0 & a_{53} & a_{54} & a_{55} & a_{56} & a_{57} & 0 & 0 & 0 & 0 \\
0 & 0 & 0 & 0 & a_{65} & a_{66} & a_{67} & 0 & 0 & 0 & 0 \\
0 & 0 & 0 & 0 & a_{75} & a_{76} & a_{77} & a_{78} & a_{79} & 0 & 0 \\
0 & 0 & 0 & 0 & 0 & 0 & a_{87} & a_{88} & a_{89} & 0 & 0 \\
0 & 0 & 0 & 0 & 0 & 0 & a_{97} & a_{98} & a_{99} & a_{910} & 0 \\
0 & 0 & 0 & 0 & 0 & 0 & 0 & 0 & a_{109} & a_{1010} & 0 \\
0 & 0 & 0 & 0 & 0 & 0 & 0 & 0 & a_{119} & a_{1110} & -1
\end{pmatrix}$$

$$A_4 = \begin{pmatrix}
-1 & b_{12} & b_{13} & 0 & 0 & 0 & 0 & 0 & 0 & 0 & 0 \\
0 & b_{22} & b_{23} & 0 & 0 & 0 & 0 & 0 & 0 & 0 & 0 \\
0 & b_{32} & b_{33} & b_{34} & b_{35} & 0 & 0 & 0 & 0 & 0 & 0 \\
0 & 0 & b_{43} & b_{44} & b_{45} & 0 & 0 & 0 & 0 & 0 & 0 \\
0 & 0 & b_{53} & b_{54} & b_{55} & b_{56} & b_{57} & 0 & 0 & 0 & 0 \\
0 & 0 & 0 & 0 & b_{65} & b_{66} & b_{67} & 0 & 0 & 0 & 0 \\
0 & 0 & 0 & 0 & b_{75} & b_{76} & b_{77} & b_{78} & b_{79} & 0 & 0 \\
0 & 0 & 0 & 0 & 0 & 0 & b_{87} & b_{88} & b_{89} & 0 & 0 \\
0 & 0 & 0 & 0 & 0 & 0 & b_{97} & b_{98} & b_{99} & b_{9101} & 0 \\
0 & 0 & 0 & 0 & 0 & 0 & 0 & 0 & b_{109} & b_{1010} & 0 \\
0 & 0 & 0 & 0 & 0 & 0 & 0 & 0 & b_{119} & b_{1110} & -1
\end{pmatrix}$$

$$A_5 = \begin{pmatrix}
-1 & c_{12} & c_{13} & 0 & 0 & 0 & 0 & 0 & 0 & 0 & 0 \\
0 & c_{22} & c_{23} & 0 & 0 & 0 & 0 & 0 & 0 & 0 & 0 \\
0 & c_{32} & c_{33} & c_{34} & c_{35} & 0 & 0 & 0 & 0 & 0 & 0 \\
0 & 0 & c_{43} & c_{44} & c_{45} & 0 & 0 & 0 & 0 & 0 & 0 \\
0 & 0 & c_{53} & c_{54} & c_{55} & c_{56} & c_{57} & 0 & 0 & 0 & 0 \\
0 & 0 & 0 & 0 & c_{65} & c_{66} & c_{67} & 0 & 0 & 0 & 0 \\
0 & 0 & 0 & 0 & c_{75} & c_{76} & c_{77} & c_{78} & c_{79} & 0 & 0 \\
0 & 0 & 0 & 0 & 0 & 0 & c_{87} & c_{88} & c_{89} & 0 & 0 \\
0 & 0 & 0 & 0 & 0 & 0 & c_{97} & c_{98} & c_{99} & c_{910} & 0 \\
0 & 0 & 0 & 0 & 0 & 0 & 0 & 0 & c_{109} & c_{1010} & 0 \\
0 & 0 & 0 & 0 & 0 & 0 & 0 & 0 & c_{119} & c_{1110} & -1
\end{pmatrix}$$

$$h_{iJ} = f_{iJ} + \delta \, D^{-1} \, m_{iJ} + \delta^2 A \, n_{iJ}$$

$$X_3 = \begin{bmatrix} w_1 \\ w_2 \\ w_3 \\ w_4 \\ w_5 \\ w_6 \\ w_7 \\ w_8 \\ w_9 \\ w_{10} \\ w_{11} \end{bmatrix} \qquad X_4 = \begin{bmatrix} \theta_1 \\ \theta_2 \\ \theta_3 \\ \theta_4 \\ \theta_5 \\ \theta_6 \\ \theta_7 \\ \theta_8 \\ \theta_9 \\ \theta_{10} \\ \theta_{11} \end{bmatrix} \qquad X_5 = \begin{bmatrix} N_1 \\ N_2 \\ N_3 \\ N_4 \\ N_5 \\ N_6 \\ N_7 \\ N_8 \\ N_9 \\ N_{10} \\ N_{11} \end{bmatrix}$$

$$B_3 = \begin{Bmatrix} e1 \\ e2 \\ e3 \\ e4 \\ e5 \\ e6 \\ e7 \\ e8 \\ e9 \\ e10 \\ e11 \end{Bmatrix} \qquad B_4 = \begin{Bmatrix} d1 \\ d2 \\ d3 \\ d4 \\ d5 \\ d6 \\ d7 \\ d8 \\ d9 \\ d10 \\ d11 \end{Bmatrix} \qquad B5 = \begin{Bmatrix} f1 \\ f2 \\ f3 \\ f4 \\ f5 \\ f6 \\ f7 \\ f8 \\ f9 \\ f10 \\ f11 \end{Bmatrix}$$

The equilibrium conditions are

$$R_3^1 + R_1^2 = 0, \qquad\qquad R_3^2 + R_1^3 = 0,$$

$$R_3^3 + R_1^4 = 0, \qquad\qquad R_3^4 + R_1^5 = 0,$$

$$Q_3^1 + Q_1^2 = 0, \qquad\qquad Q_3^2 + Q_1^3 = 0,$$

$$S_3^1 + S_1^2 = 0, \qquad\qquad S_3^2 + S_1^3 = 0,$$

$$S_3^3 + S_1^4 = 0, \qquad\qquad S_3^4 + S_1^5 = 0, \qquad\qquad 4.4$$

Solving these couples global matrices for temperature, concentration and velocity (4.1) – (4.4) respectively and using the iteration procedure we determine the unknown global nodes through which the temperature, concentration and velocity at different radial intervals at any arbitrary axial cross sections are obtained. The respective expressions are given by

$$\theta(r) = \psi_1^1\theta_{11} + \psi_2^1\theta_{12} + \psi_3^1\theta_{13} \qquad\qquad 1 \le r \le 1 + S^*0.2$$

$$= \psi_1^2\theta_{13} + \psi_2^2\theta_{14} + \psi_3^2\theta_{15} \qquad\qquad 1 + S^*0.2 \le r \le 1 + S^*0.4$$

$$= \psi_1^3\theta_{15} + \psi_2^3\theta_{16} + \psi_3^3\theta_{17} \qquad\qquad 1 + S^*0.4 \le r \le 1 + S^*0.6$$

$$= \psi_1^4\theta_{17} + \psi_2^4\theta_{18} + \psi_3^4\theta_{19} \qquad\qquad 1 + S^*0.6 \le r \le 1 + S^*0.8$$

$$= \psi_1^{15}\theta_{19} + \psi_2^5\theta_{20} + \psi_3^5\theta_{21} \qquad\qquad 1 + S^*0.8 \le r \le 1 + S^*$$

$$N(r) = \psi_1^1 N_{11} + \psi_2^1 N_{12} + \psi_3^1 N_{13} \qquad\qquad 1 \le r \le 1 + S^*0.2$$

$$= \psi_1^2 N_{13} + \psi_2^2 N_{14} + \psi_3^2 N_{15} \qquad\qquad 1 + S^*0.2 \le r \le 1 + S^*0.4$$

$$= \psi_1^3 N_{15} + \psi_2^3 N_{16} + \psi_3^3 N_{17} \qquad\qquad 1 + S^*0.4 \le r \le 1 + S^*0.6$$

$$= \psi_1^4 N_{17} + \psi_2^4 N_{18} + \psi_3^4 N_{19} \qquad\qquad 1 + S^*0.6 \le r \le 1 + S^*0.8$$

$$= \psi_1^{15} N_{19} + \psi_2^5 N_{20} + \psi_3^5 N_{21} \qquad\qquad 1+S^*0.8 \leq r \leq 1+S$$

$$w(r) = \psi_1^1 w_{11} + \psi_2^1 w_{12} + \psi_3^1 w_{13} \qquad\qquad 1 \leq r \leq 1+S^*0.2$$

$$= \psi_1^2 w_{13} + \psi_2^2 w_{14} + \psi_3^2 w_{15} \qquad\qquad 1+S^*0.2 \leq r \leq 1+S^*0.4$$

$$= \psi_1^3 w_{15} + \psi_2^3 w_{16} + \psi_3^3 w_{17} \qquad\qquad 1+S^*0.4 \leq r \leq 1+S^*0.6$$

$$= \psi_1^4 w_{17} + \psi_2^4 w_{18} + \psi_3^4 w_{19} \qquad\qquad 1+S^*0.6 \leq r \leq 1+S^*0.8$$

$$= \psi_1^{15} w_{19} + \psi_2^5 w_{20} + \psi_3^5 w_{21} \qquad\qquad 1+S^*0.8 \leq r \leq 1+S$$

The shear stress (τ) is evaluated using the formula $\tau = \left(\dfrac{du}{dr} \right)_{r=1,S}$

The rate of heat transfer (Nusselt number) is evaluated using the formula

$$Nu = -\left(\frac{d\theta}{dr} \right)_{r=1,S}$$

The couple stress is evaluated using formula

$$N = -\left(\frac{dN}{dr} \right)_{r=1,S}$$

3.5. DISCUSSION:

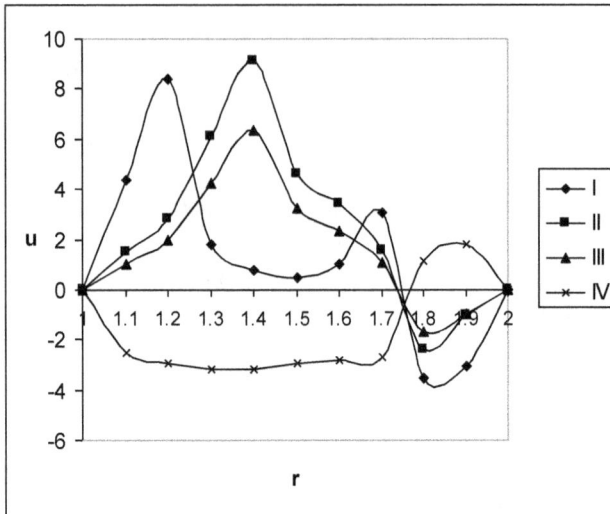

Fig. 1 : Variation of velocity (u) with G

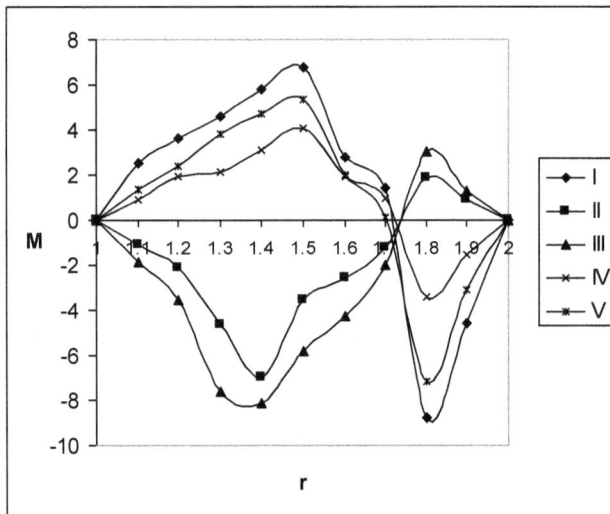

Fig. 2 : Variation of velocity (u) with G

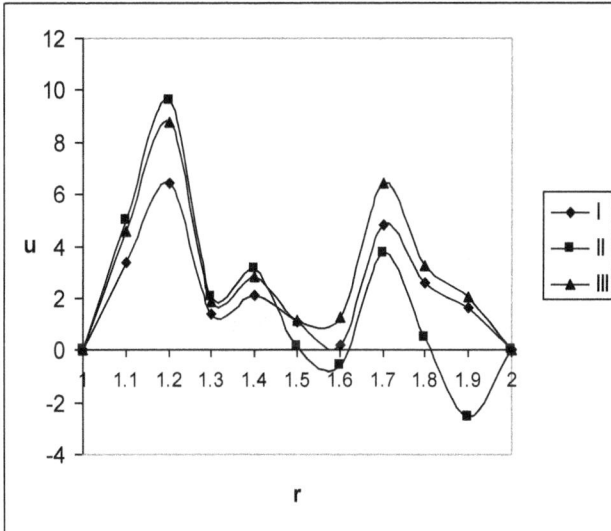

Fig. 3 : Variation of velocity (u) with Δ

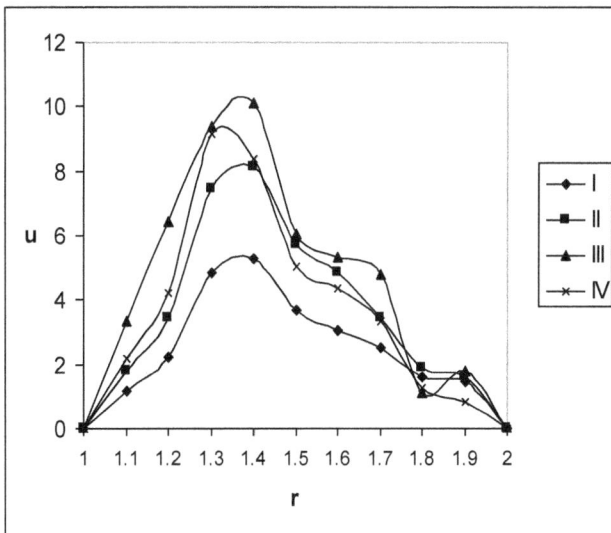

Fig. 4 : Variation of velocity (u) with λ

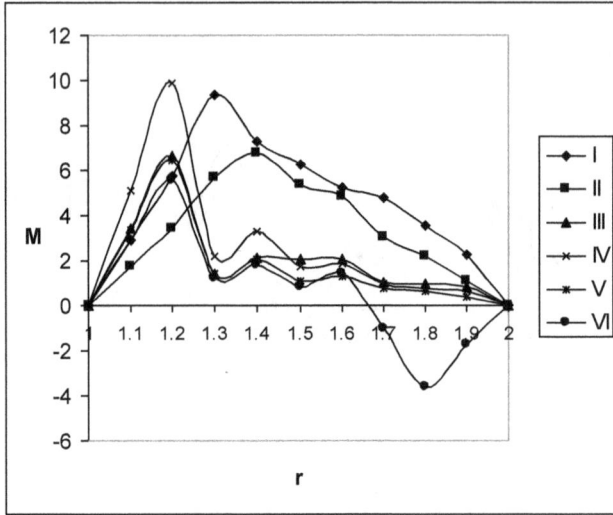

Fig. 5 : Variation of velocity (u) with S

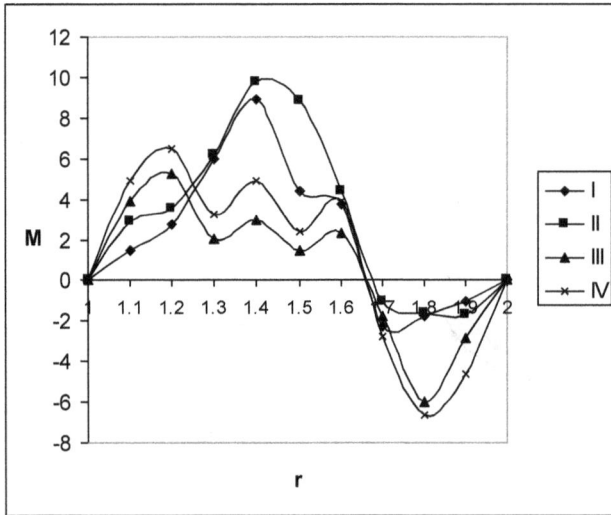

Fig. 6 : Variation of velocity (u) with N_t

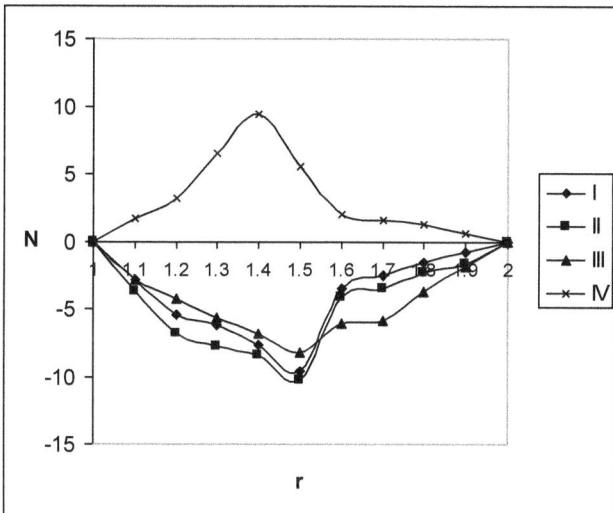

Fig. 7 : Variation of micro rotation (N) with G

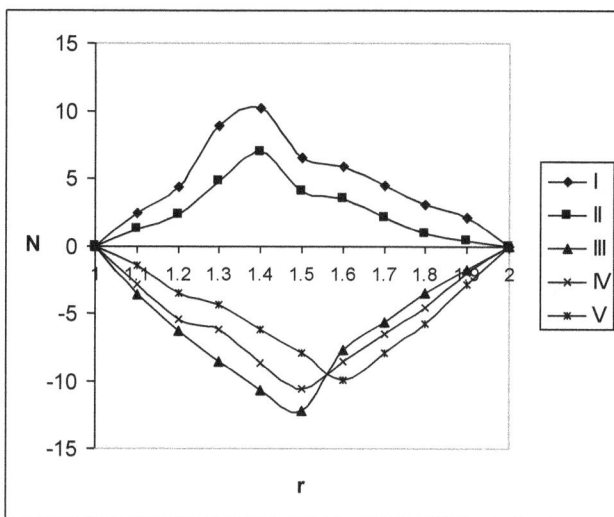

Fig. 8 : Variation of micro rotation (N) with G

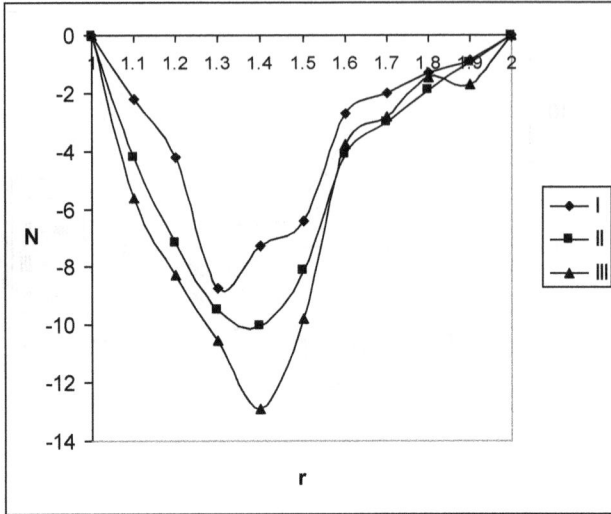

Fig. 9 : Variation of micro rotation (N) with Δ

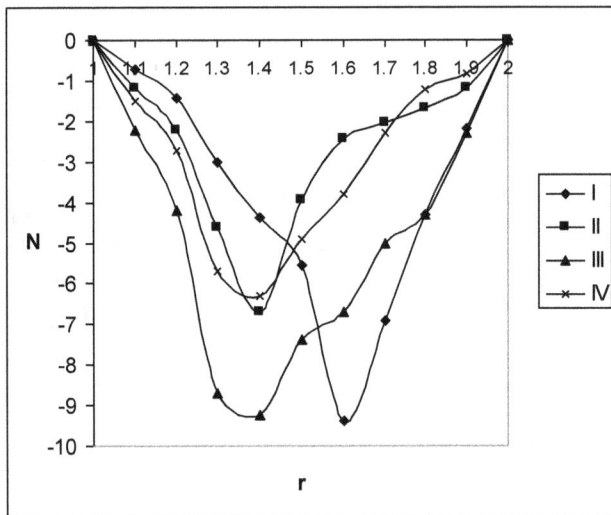

Fig. 10 : Variation of micro rotation (N) with λ

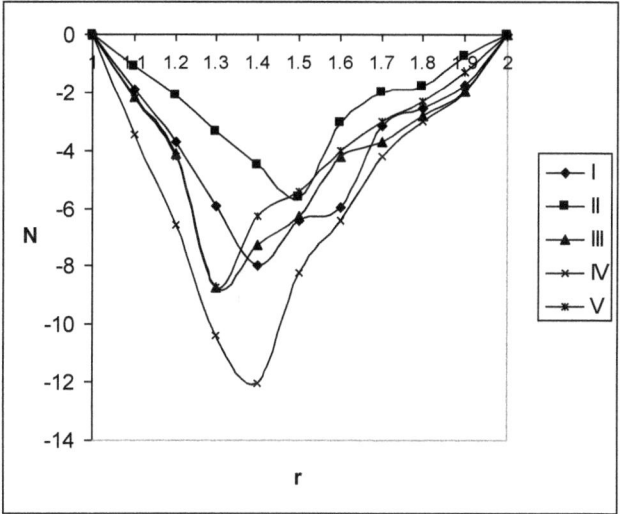

Fig. 11 : Variation of micro rotation (N) with S

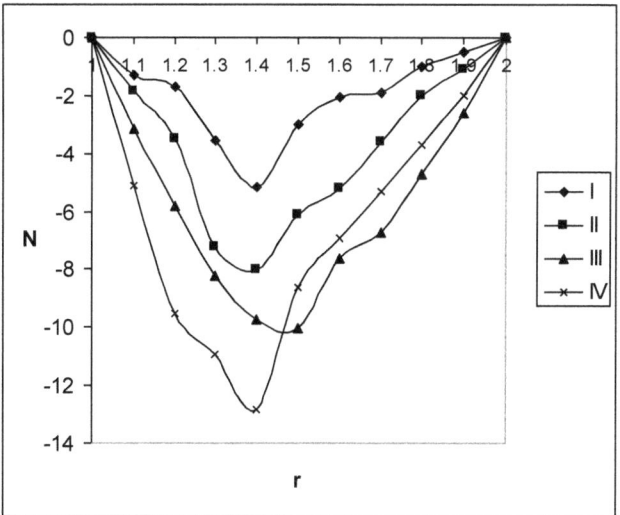

Fig. 12 : Variation of micro rotation (N) with N_t

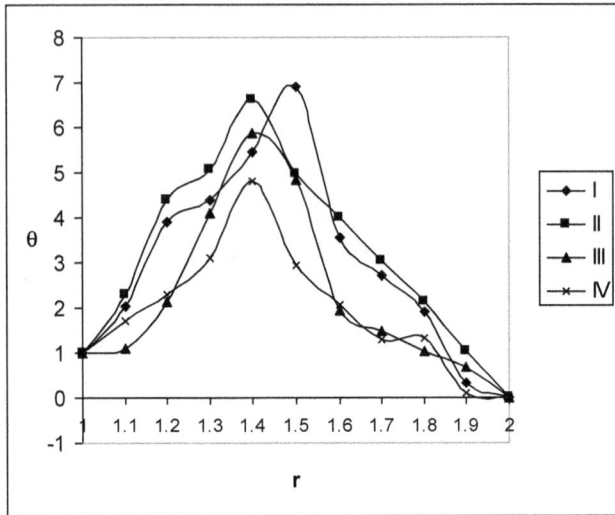

Fig. 13 : Variation of temperature (θ) with G

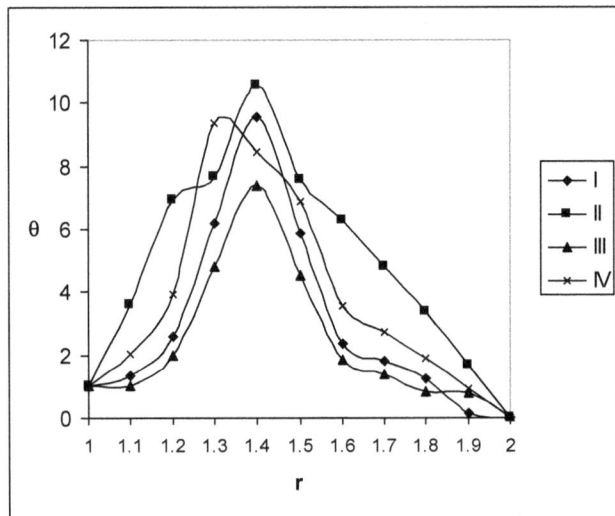

Fig. 14 : Variation of temperature (θ) with G

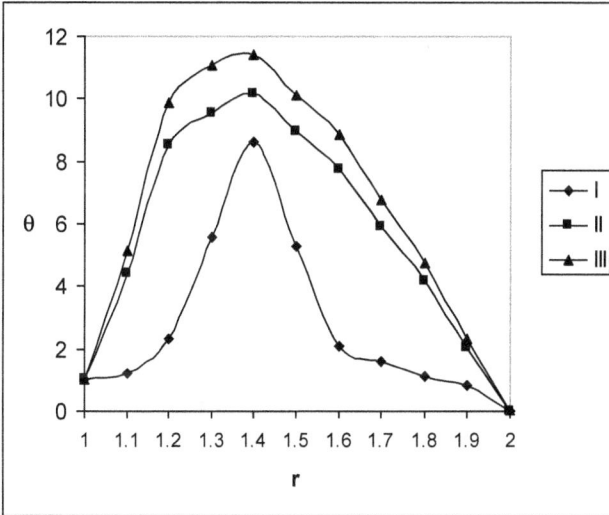

Fig. 15 : Variation of temperature (θ) with Δ

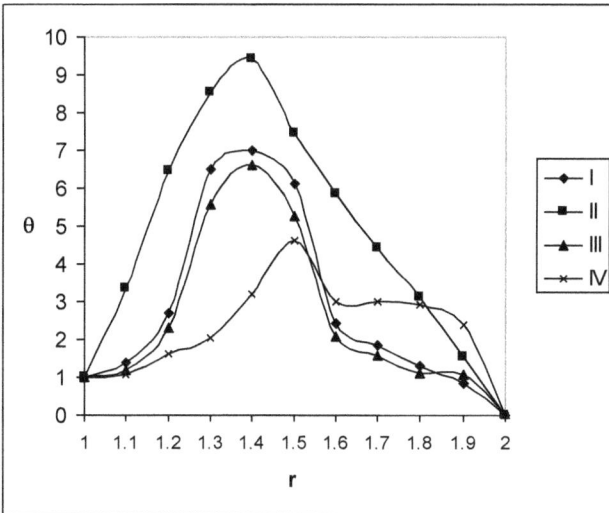

Fig. 16 : Variation of temperature (θ) with λ

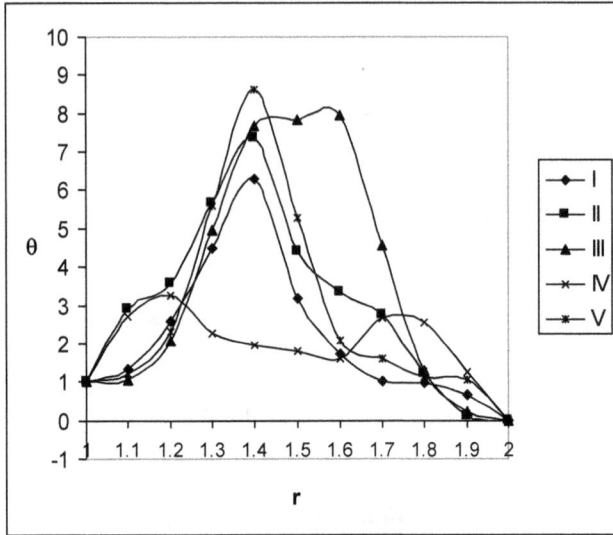

Fig. 17 : Variation of temperature (θ) with S

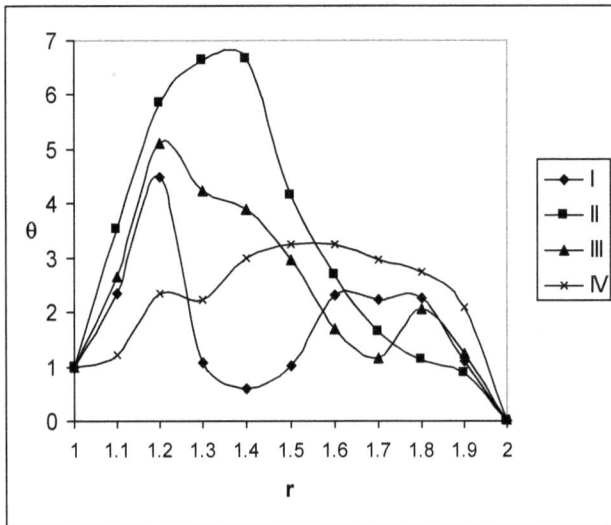

Fig. 18 : Variation of temperature (θ) with N_t

88

The equations, given by the finite element technique the velocity, micro rotation and temperature distributions are obtained. The Prandtl number Pr , material constants A & A are taken to be constant, at 0.733, 1 and 1 respectively whereas the effect of other important parameters, namely micro polar parameter Δ, the section Reynolds number λ, Grashof number G and Darcy parameter D^{-1} has been studied for these functions and the corresponding profiles are shown in figs. 1.

Fig. 1 depicts the variation of velocity function W with Grashof number G. The actual axial velocity W is in the vertically downwards direction and w < 0 represents the actual flow. Therefore w > 0 represents the reversal flow. We notice from fig. 1 that w>0 for G > 0 and w < 0 for G < 0 except in the vicinity of outer cylinder r = 2. the reversal flow exists everywhere in the region (1.1 $\leq$ r $\leq$ 1.8) for G > 0 and in the neighborhood of r = 2 for G < 0. The region of reversed flow enlarges with G $\leq$ 2x10^3 and shrinks with higher G $\geq$ 3x10^3. Also it grows in size with |G| (<0). |w| enhances with G $\leq$ 2x10^3 and reduces with higher G $\geq$ 3x10^3. |w| experiences a depreciation in the case of cooling of the boundaries with maximum at r = 1.5. The variation of w with Darcy parameter D^{-1}. It is found that lesser the permeability of the porous medium larger |w| in the flow region (fig. 2). The influence of micro rotation parameter Δ on w is shown in fig. 3. As the micro polar parameter Δ $\leq$ 3 increases, the velocity continuously increases and decreases with higher Δ $\geq$ 5, with maximum attained in the vicinity of r = 1. Fig. 4 represents w with suction parameter λ. IT is found that the axial velocity experiences an enhancement with increase in λ $\leq$ 0.03 and depreciates with higher λ $\geq$ 0.05. Fig. 5 depicts w with the width of the annular region. We notice that the axial velocity continuously decreases with increase in the width S of the annular region. Thus the velocity enhances in the narrow gap region and depreciates in the wide gap case. The effect of radiation parameter N_1 on w is exhibited in Fig. 6. IT is observed that the velocity w enhances with increase in $N_1 \leq 1.0$ except in a narrow region adjacent to $r = 2$ and for higher $N_1 \geq 1.5$, it experiences an enhancement in the entire flow region.

The micro rotation (N) in shown in figs 7-12 for different values of G, Δ, λ, S and N_1. It is found that the values of micro rotation for $G > 0$ are negative and positive for $G < 0$. An increase in $|G| \leq 2 \times 10^3$ editor an enhancement in N and for higher $|G| \geq 3 \times 10^3$, it reduces in the region adjacent to $r = 1$ and enhances in the reform adjacent to $r = 2$ with maximum at $r = 1.5$ (fig. 7). From fig. 8 we find that lesser the permeability of the porous medium larger the micro rotation everywhere in the flow region. The effect of N on micro polar parameter Δ in shown in Fig. 9. We notice that an increase in $\Delta \leq 3$ leads to an enhancement in $|N|$ and for higher $\Delta \geq 5$, it enhances the first half ($1.1 \leq r \leq 1.5$) and reduces in the second half ($1.6 \leq r \leq 1.9$). Fig. 10 illustrates that the micro rotation enhances in the first half and reduces in the second half with increase in $\lambda \leq 0.02$ and for higher values of $\lambda \geq 0.03$ we notice an increment in $|N|$ everywhere in the flow region. Form fig. 11 we find that the micro rotation depreciates in the narrow gap case and enhances in the wide gap case. Fig. 12 illustrates that the micro rotation $|N|$ enhances with increase in the radiation parameter $N_1 \leq 1.5$ and for higher $N_1 \geq 2.5$, the micro rotation enhances in the let half and depreciates in the second half of the flow region.

The non-dimensional temperature (θ) is shown in figs. 13 – 17 for different values of G, λ, Δ, S and N_1. Fig. 13 illustrates that non-dimensional temperature is positive for all variations. The actual temperature enhances with increase in $G \leq 2 \times 10^3$ and depreciates with higher $G \geq 3 \times 10^3$. Also it enhances with $G < 0$. The variation of θ with D^{-1} shows that lesser the permeability of the porous medium larger the actual temperature in the flow region (Fig. 14), Fig. 15 illustrates that an increase in the micro rotation parameter Δ increases the actual temperature continuously with maximum attained at $r=1.5$. The variation of θ with suction parameter λ shows that the temperature enhances with increase in $\lambda \geq 0.03$. For further increase in $\lambda \geq 0.05$, the temperature depreciates in the first half and enhances in the second half. The influence of the suction of the boundary on θ is shown in fig. 16. IT is found that increase in $S \leq 0.6$ reduces θ in the left half and enhances it in the right half and for $S = 0.7$, θ reduces in the flow region except in a region adjacent to $r = 1$. For further increase in S we notice an enhancement in the central flow region and depreciation in θ in the regions abutting the cylinders $r = 1$ & 2. Fig. 17 represents the temperature with radiation parameter N_1. It is found that an increase in N_1 enhances θ in the left half and depreciates in the right half and for higher values $N_1 \geq 1.5$ we observe depreciation in the first half and enhancement in the second half.

The shear stress (τ) at the inner and outer cylinders $r = 1$ & 2 are evaluated for different G, D^{-1}, Δ, λ and S and are exhibited in tables 1-4. It is found that the stress at the boundaries is positive for $G > 0$ and negative for $G < 0$. From the numerical results it is noticed that the stress (τ) enhances at $r = 1$ and depreciates at $r = 2$ with increase in $|G| \geq 0$. The variation of τ with D^{-1} shows that lesser the permeability of the porous medium larger τ in the heating case and smaller $|\tau|$ in the cooling case at $r = 1$ and at $r = 2$ larger $|\tau|$ for all G. An increase in the suction parameter $\lambda \leq 0.03$ enhances $|\tau|$ and for higher $\lambda \geq 0.05$ smaller $|\tau|$ for $G > 0$ and larger $|\tau|$ for $G < 0$ at $r = 1$. While at $r = 2$, smaller $|\tau|$ $\lambda \leq 0.03$ and larger $|\tau|\lambda \leq 0.05$ for $G > 0$. the variation of τ with micro polar parameter Δ shows that an increase $\Delta \leq 3$ enhance $|\tau|$ and reduces with higher $\Delta \geq 5$ $|G|=10^3$ while a reversed effect is noticed at $|G| = 3 \times 10^3$ at the cylinders $= 1$ & 2. At $r = 2$, it enhances with Δ at $|G| = 10^3$ fixing the other parameter G and λ. The variation of τ with annular width is reveals that stress at $r = 1$, depreciates with increase in S at $|G| = 10^3$ and enhances at $|G| = 3 \times 10^3$. At $r = 2$, it enhances with $S \leq 0.6$ and depreciates with $G \geq 0.8$ at $|G| = 10^3$ and at $|G| = 3 \times 10^3$, it depreciates for all values S. (table. 4).

The Nusselt number (Nu) which measures the rate of heat transfer at $r = 1$ & 2 are shown in tables 5-8 for different parametric values. It is found that the rate of heat transfer depreciates at r = 1 and enhances at r = 2 with increase in Grashof number G. The variation of Nu with D^{-1} reveals that lesser the permeability of the porous medium larger Nu at r = 1 & 2 for G < 0 while for G > 0, it enhances at r = 2 and at r = 1, it reduces with $D^{-1} \leq 2 \times 10^2$ and enhances with higher D^{-1} 3×10^2. An increase in the suction parameter $\lambda \leq 0.03$ reduces |Nu| and enhances with higher $\lambda \geq 0.05$ at r = 1 & 2 for $|G| \geq 10^3$ and at $|G| \geq 3 \times 10^3$, higher |Nu| at both the cylinders. The variation of Nu with Δ shows that the rate of heat transfer at r = 1 enhances with $\Delta \leq 3$ and reduces with higher $\Delta \geq 5$ at $|G| = 10^3$ while a effect is found at $|G| = 10^3$. At r = 2, |Nu| enhances with Δ for G > 0 and reduces for G < 0. The influence of annular widths on Nu at r = 1 & 2 is shown in tables 6 & 8. It is found that the rate of heat transfer experiences an enhancement with increase in S for all G at both the boundaries. This implies that the micro polar parameter Δ are well as Darcy parameter D^{-1} are the important parameters in controlling the rate of heat transfer, which is desired in many technological applications.

The couple stress (M*) at the inner and outer cylinders r = 1 & 2 are exhibited in tables 9-12 for different parametric values. It is found that an increase in |G| reduces M* at r = 1 and enhances at r =2. Lesser the permeability of the porous medium larger M* at r = 1 and at r = 2 larger M* and for further lowering of the permeability smaller M* for al G (tables 9). An increase in $\lambda \leq 0.03$ reduces and enhances with higher $\lambda \geq 0.05$ in the heating case and a reversed effects observed in the cooling case at both the cylinders. The variation of M* with micro polar parameter Δ shows that an increase in $\Delta \leq 3$ enhances M* at r = 1 and reduces at r = 2 and for higher $\Delta \geq 5$, it reduces at r = 1 and enhances at $|G| = 10^3$. AT higher $|G| = 3 \times 10^3$, M* reduces with Δ and enhances with $\Delta \geq 5$ at r = 1 (tables (11 & 2). Also the couple stress at r = 1 reduces with $S \leq 0.6$ and enhances with higher $S \geq 0.8$ for G > 0 and for G<0, a reversed effects noticed. AT r = 2, it enhances with $S \leq 0.6$ and reduces with $S \geq 0.8$.

REFERENCE:

[1]   T. ARIMAN, J. Biomech. 4. 185 (1971).

[2]   T. ARIMAN, M.A, TURK and N.D. SYLVESTER, Int. J. Eng, sci. 11, 905 (1973).

[3]   R.S. AGARWAL[1] and C. DHANAPAL[2] Numerical solution of micro polar fluid flow and heat transfer between two co-axial porous circular cylinders Int. J. Eng. Sci, vol. 26. No.11, pp. 1133-1142, 1988.

[4]   A.S. BERMAN, J. appal. Phys, 29, 71 (1958).

[5]   A.C. ERINGEN, J. Math. Mech. 16. 1 (1966).

[6]   A.C. ERINGEN, J. Math. Analysis Applic. 38 480 (1972).

[7]   HAVSTAD, M.A, BURNS, P.J. convective heat transfer in vertical cylindrical Annuli filled with porous medium, Int. J. Heat Mass Transfer. 25 (1982), 11, pp. 1755-1766.

[8]   HICKON, C.E, GARTLING, D.K. A Numerical study of natural convection in a vertical Annulus porous layer, Int. J. Heat Mass Transfer, 28 (1985) 3, pp. 720-723.

[9]   G.R. INGER, phys. Fluids 12, 1741 (1969).

[10] Y. KAIZAKI and T. ARIMAN, Rheol. Acta 10 319 (1971).

[11] S.P. MISHRA and B.P. ACHARYA. Ind. J. phys. 46, 469 (1972).

[12] PRASAD, V., KULACKI, F.A. Natural convection in a vertical porous Annulus, Int. J. Heat Mass Transfer 27 (1984), 2 pp. 207-219.

[13]   PRASAD, V. KULACKI, F.A. Natural convection in porous media bounded by short concentric cylinders, ASME J. Heat Transfer, 107 (1985), 1. pp. 147-154.

[14]   PRASAD, V., KULACKI, F.A, KEYHANI, M. Natural convection in porous media, J. Fluid Mechanics, 150 (1985), 3, pp. 89-119.

[15]   REDA, D.C. Natural convection experiments in a liquid saturated porous medium bounded by vertical coaxial cylinders, ASME J. Heat Transfer, 105 (1983), 4, pp. 795-802.

[16]   SHIVAKUMARA, I.S, PRASANNA, B.M.R, RUDRAISH, N. VENKATACHALAPPA, M. Numerical study of natural convection in a vertical cylindrical Annulus using Non-Darcy equation. J. Porous Media, 5 (2003), 2, pp. 87-102.

[17]   N.D. SYLVESTER, M.A, TURK and T. ARIMAN Trans. Soc. Rheol. 17, 1 (1973).

[18]   SASTRY, V.U.K, MAIT, G: Numerical solution of combined convective heat transfer of micro polar fluid past a non-isothermal vertical flat plate.

ABOUT AUTHORS

Dr. B. Tulasi Lakshmi Devi is a working as Associate Professor of Mathematics, Guru Nanak Institute of Technology Hyderabad. She obtained Ph.D. Degree from Sri Krishnadevaraya University, Anantapur. Her field of interest research area is Convective Heat Transfer .She has published several articles in various journals to her credit. She has more than 10 years of teaching experience of engineering colleges. She presented around six papers in various conferences. She is a member of professional Bodies of APSMS, ISTE.

Dr.G.Srinivas working as Professor in the Department of Mathematics, Guru Nanak Institute of Technology, Hyderabad. He received his Ph.D from Sri Krishnadevaraya University, Anantapur in 2006. In the field of interest includes Fluid Dynamics, Heat and Mass Transfer. He is Mathematician, philanthropist. He made Numerous research presentations, organized and contributed paper sessions, and served as the reviewer at National and International conferences in Mathematics. He has supervised Research scholars for their Ph.D. His focus is helping, developing and implement a new teaching and learning framework.

G.V.P.N.Srikanth, Research Scholar in JNTUH, Hyderabad. He received M.Sc degree in the stream of Mathematics from Andhra University in 2007, and pursuing Ph.D in the stream of convective Heat and Mass Transfer from JNTUH, Hyderabad. He has presented Research papers in National and International conferences and also published papers in International Journals. In the field of interest includes Fluid Dynamics, Heat and Mass Transfer. He is a Mathematician and logician. He received numerous honors and awards including the best Teacher award.

www.ingramcontent.com/pod-product-compliance
Lightning Source LLC
Chambersburg PA
CBHW060632210326
41520CB00010B/1574